Adult literacy in Britain

A survey of adults aged 16-65 in Great Britain carried out by Social Survey Division of ONS, and commissioned by a consortium of Government Departments and the Basic Skills Agency, the main funding being provided by the Department for Education and Employment.

This survey was carried out as part of the International Adult Literacy Survey (IALS)

Siobhán Carey

Sampson Low

Jacqui Hansbro

London: The Stationery Office

© Crown copyright 1997

First published 1997

ISBN 0 11 620943 7

Designed by Adkins Design

Cover: Library photographs

Contents

Authors' acknowledgments

Any large scale survey involves the co-operation and collaboration of a large number of people. The Adult Literacy Survey was the first time Social Survey Division undertook this type of survey which fuses educational and social research. It was an interesting and rewarding experience for all concerned. This survey placed a particularly heavy burden on the respondent who was asked to complete a lengthy selection of tasks as well as the interview. We are particularly indebted to the many people who gave up their time to take part in this study. As the literacy assessment was not subject to any time limits it presented additional difficulties for interviewers in planning their work. We would like to thank the interviewers for all the extra work that this survey required, for the late nights they spent out in inclement weather and for their feedback on progress which kept us reassured. We hope that both the respondents and the interviewers found it an enjoyable experience.

This survey in particular involved the close collaboration of colleagues in many other countries. All the countries taking part in the International Adult Literacy Survey worked closely together to achieve a common goal. It is not often that we get to work so closely with survey organisations in other countries in implementing a survey. We would like to thank those involved at Statistics Canada and Educational Testing Service for their help and guidance during the project and for their willingness to answer what must have sometimes seemed like an endless series of questions. We would also like to thank the national study teams in the other countries for making this study such a rewarding experience.

The project team in ONS comprised

Research	Siobhán Carey *(Project Manager)*
	Sampson Low
	Jacqui Hansbro
Fieldwork management	Marily Troyano
	Diane Worthington
Computing support	Gavin Cotgrove
Scoring/Administration	Rebecca Sale
	Eimear Schlindwein
Group Director	Joy Dobbs

Notes to tables

1. The following conventions have been used in tables:

-        No cases

0   Values of less than 0.5%

[ ]   the numbers in square brackets are the actual numbers which are presented when the base is less than 30.

*   denotes than the mean is not shown as the base was less than 65

..   category not applicable

2. The row percentages may add to 99% or 101% because of rounding.

3. Unless stated otherwise, changes and differences mentioned in the text have been found to be statistically significant at the 95% confidence level.

4. Classification variables such as economic activity status and occupation are not presented in a standard form throughout the report partly because of differences between standard classifications used in Britain and those required by the international context of the survey but also because of the limitations of the sample size. For some analyses, categories had to be grouped together in order to ensure sufficient cases for analysis. This mostly occurred where there was a high level of association between literacy and the variable(s) of interest. For example, on some analyses the classification of educational attainment level and occupational group had to be collapsed because of the homogeneity of some occupations in terms of educational attainment.

5. Where there is a percentage sign at the head of a column the whole distribution is presented and the figures add to 100% (see note 2). Where there is no percentage sign in the table and a note above the figures, the figures refer to the proportion of people who had the attribute being discussed.

Summary of main findings

Chapter 1 Introduction

The Adult Literacy survey carried out in Britain in 1996 forms part of an international programme of surveys known as the International Adult Literacy Survey (IALS). By the end of 1998 over 20 countries will have participated in the study worldwide. The British Survey was carried out by the Social Survey Division of the Office for National Statistics and was commissioned by a consortium of Government Departments and the Basic Skills Agency, the main funding being provided by the Department for Education and Employment.

This survey is the first literacy survey to be carried out in Britain on a national random probability sample of adults of working age. It set out to profile the literacy abilities of adults aged 16-65 using an internationally agreed measurement instrument and internationally agreed survey implementation protocols covering such aspects as interviewer instructions and scoring procedures.

The definition of literacy used in IALS is:

Using printed and written information to function in society, to achieve one's goals and to develop one's knowledge and potential.

This definition does not treat literacy as a dichotomous condition that people either have or do not have, but rather defines literacy as a broad range of skills required in a varied range of contexts. The survey measured three dimensions of literacy:

- **Prose literacy:** the knowledge and skills required to understand and use information from texts such as newspaper articles and passages of fiction.

- **Document literacy:** the knowledge and skills required to locate and use information contained in various formats such as timetables, graphs, charts and forms.

- **Quantitative literacy:** the knowledge and skills required to apply arithmetic operations, either alone or sequentially, to numbers embedded in printed materials, such as calculating savings from a sale advertisement or working out the interest required to achieve a desired return on an investment.

Performance on each of these scales has been grouped into five literacy levels; Level 1 represents the lowest ability range and Level 5 the highest. Because of the small proportion of people at the highest level, Level 5 , data are presented for Levels 4 and 5 combined (Level 4/5).

Chapter 2 Distribution of literacy skills

- 22% of the population of working age performed at Level 1 on the prose scale, 30% at Level 2, 31% at Level 3 and 17% at Level 4/5. The distribution was similar on the document and quantitative scales.
- Among men, performance on the prose scale was poorer than on either the document or quantitative scales whereas women performed better on the prose scale. Significantly higher proportions of women than men were at Level 1 on both document and quantitative literacy (27% and 29% of women compared with 20% and 18% of men).
- On all three scales there was a higher proportion performing at the lowest literacy level, Level 1, in the two oldest age-groups, that is, those aged over 45 than in the younger age-groups.
- The distribution of literacy levels for the three youngest age-groups, 16-25, 26-35 and 36-45 was very similar.

- Literacy was strongly associated with education, the percentage of people performing at the higher literacy levels increasing with increasing education. Those with lower levels of education were much more likely to be at literacy Levels 1 and 2.

- Education was not always a good predictor of literacy, however, among those with higher education there was a small proportion of individuals who performed at Level 1 and among those with primary education or lower there was a small proportion who were at Level 4/5.

- Those in employment and full-time students were more likely than the unemployed or economically inactive to perform at the highest literacy levels, Level 4/5, on all three scales. The unemployed were twice as likely as those in employment to perform at Level 1.

- On all three literacy scales those in manual social class groups were much more likely than others to be at the lower literacy levels. Over a third (36%) of those in Social Classes IV and V and 29% of those in Social Classes III (manual) were at Level 1 on prose compared with 12% or less for other social class groups.

- Those in the lowest income quintile were more likely to perform at Level 1 than those in the two highest income quintiles.

- Respondents in receipt of social security benefits (excluding pensions and child benefit) were much more likely to have low literacy skills than those not in receipt of benefits.

- There was no difference in the distribution of literacy levels for England and Scotland on any of the three literacy scales but there were significant differences between England and Wales with Wales having a smaller proportion of adults at Level 4/5 (9% on the prose scale for example compared with 17% in England).

Chapter 3 Literacy and work

- The industries that have seen the greatest increase in the number of employees over recent years, such as the financial sector and the sector that includes information technology and research, have larger proportions of employees at the higher literacy levels.

- Employees in industries that have seen the greatest decline in employment, such as agriculture/mining or construction, were more likely to be at the lower literacy levels than employees in most other sectors.

- Those in managerial, professional or technical occupations were more likely to perform at the higher literacy levels than those in other occupations. About a third of associate professional and technical occupations performed at Level 4/5 on prose literacy with higher proportions performing at that level for both the document and quantitative dimensions.

- Workers in clerical and secretarial occupations were more evenly distributed over the literacy skill levels with some workers at every level, although few (11% - 14%) were at Level 1.

- Workers in personal and protective service, sales and skilled engineering occupations were more likely to perform at the middle literacy levels, Levels 2 or 3, while those in occupations such as machine operators and other elementary occupations were more likely to be at the two lowest literacy levels, Levels 1 and 2.

- Respondents were asked how often they had to perform different types of reading, writing and mathematical activities as part of their job. Managers, professionals and associate professionals were all regularly required to perform most of the reading tasks in contrast to the relatively low demand for workplace reading in some other occupations such as plant and machine operative and other elementary occupations.

- Even in those occupations with poor average proficiency levels on the three literacy dimensions, substantial proportions (a third to a half) of workers were regularly required to undertake activities that required reading skills.

- Generally, those who reported engaging in literacy activities at least once a week as part of their job demonstrated higher literacy skills on each dimension than those who reported engaging in these activities less frequently. This does not necessarily imply that engaging in these activities at work leads to high literacy as large proportions of those with low skill levels are also required to undertake literacy tasks as part of their job.

- Overall very few people rated either their reading, writing or mathematical skills as poor for their job. People generally rated their reading skills higher than either their writing or mathematics skills.

- Only a small proportion of respondents (2% - 15%) felt that their reading, writing or mathematics skills were limiting their job opportunities - either job advancement or getting another job. Those at the lower literacy levels were more likely to consider their skills as limiting their opportunities.

- Over half (57%) of those who had worked in the 12 months prior to interview had participated in some form of education or training during the previous the 12 month period but those in occupations with low average literacy skill were least likely to have received any education or training in the reference period.

- Those with the greatest income from employment were also those with the highest literacy skills; almost half (47%) of those in the highest income quintile performed at level 4/5 on quantitative literacy and a further 36% were at level 3 with very few demonstrating low quantitative literacy ability.

Chapter 4 Literacy in everyday life

- Those who read books daily were more likely than those who never read books to be at Levels 3 or above on all three literacy dimensions.

- Generally, respondents were very pleased with their literacy skills in the context of their daily lives, with 46% considering their reading skills as 'excellent' and a further 40% describing them as 'good' even though 52% of all respondents were at the lowest two levels on the prose scale.

- Although a small proportion of respondents assessed their reading, writing or mathematics skills for daily life as poor, on all three dimensions the majority of those that did so were at Level 1.

- Respondents at Level 1 were less likely to report that their children would often see the respondent or their partner reading than were those at higher literacy levels. Almost all respondents at prose Level 4/5 said their children would often see them or their partner reading.

Chapter 5 People with low literacy skills

- People at each end of the literacy distribution - those with low literacy skills (Level 1) and those with high literacy skills (Level 4/5) - each form a clearly defined, relatively homogenous group while people at Levels 2 and 3 are more diverse in their characteristics and are less clearly defined as a group.

- People performing at Level 1 on the literacy scales were predominantly older people with low levels of education. They were more likely than people at higher levels to be unemployed, to belong to the manual rather than the non-manual social classes and to be on a low income. Not all those with these characteristics, however, will have low literacy skills.

- Those at Level 4/5 were predominantly young (aged 45 or under), with high levels of education, although a notable minority (29% on the prose scale) had not continued their formal education beyond lower secondary level. Those at Level 4/5 on the document and quantitative scales were more likely to be men. People at Level 4/5 were the most likely to be in employment and to be in non-manual social classes.

- On each of the scales, those at Level 1 were less likely than those at Level 4/5 to say they read a book at least once a week (10% compared with 23% on the prose scale).

- Among those who performed at the lowest literacy level a large proportion considered their skills to be adequate for daily life; around three quarters were 'very satisfied' or 'somewhat satisfied' with their reading and writing skills.

- Of those performing at Level 1 on quantitative literacy, 18% recognized that their mathematical ability was poor and a further 45% considered it moderate.
- Although people at Level 1 were on the whole satisfied with their literacy skills they were more likely than those at the higher levels to say that they sometimes or often needed help with various literacy tasks, particularly filling out forms and reading information from government departments, businesses and other institutions (almost half sometimes or often needed help with one or more tasks).

Chapter 6 Literacy skills in other countries

One of the objectives of IALS was to examine the determinants of literacy across a number of countries, languages and cultures. Despite every effort to ensure uniformity in survey design and implementation there are differences between the countries in response rates and sample design. It is recommended, therefore, that comparisons of results across different countries should be undertaken with due caution. The results from the British survey were compared with those from the countries that took part in the first round of IALS in 1994; Germany, Sweden, the Netherlands, Switzerland, Poland , the USA and Canada.

- The distribution of prose, document and quantitative literacy skills differ between countries. Britain's distribution was similar to that in the other English speaking countries, the United States and Canada, with slightly higher proportions at Levels 2 and 3 than there are at Levels 1 and Level 4/5.
- Relative to other countries, Britain and the United appear more polarised with relatively large proportions of the population at both the lower and upper literacy levels.
- In Britain, as in all countries except the USA the proportion of the population at literacy Levels 1 and 2 is highest in the oldest age-groups. In the USA there was also a high proportion of young people at these levels.
- In all countries people with higher levels of educational attainment tended to perform at higher literacy levels on all three scales but the relationship is stronger is some countries than in others. Among those with lower secondary and upper secondary education only Sweden and Germany had significantly lower proportions of people than Britain at the lowest literacy level, Level 1, on prose.

- In all countries apart from the French speaking part of Switzerland people performing at the lower literacy levels were more likely to be unemployed than those who performed at the higher literacy levels.
- In Britain and in all other countries the expected relationship between literacy and occupation was observed with large proportions of managers/professionals and technicians performing at the higher literacy levels on all three scales.
- The proportion of the population who reported engaging in the various reading tasks was very similar in the three English speaking countries, Canada, the US and Britain.
- On all reading activities a higher proportion of people in Sweden than in Britain engaged in them at least once a week as part of their job. This may contribute to the better performance of the Swedish respondents on the literacy measures.

1 Introduction

1.1 Introduction

This report presents the findings of the Adult Literacy Survey carried out in Great Britain in 1996. The survey forms part of an international programme of surveys known as the International Adult Literacy Survey (IALS). Seven countries carried out the survey in 1994 (Germany, Sweden, the Netherlands, Switzerland, Poland, the USA and Canada) and the results were published in 1995.[1] A second round of data collection was carried out in 1996 involving four countries, the UK, Australia, New Zealand and Belgium (Flemish community). A number of other countries are currently undertaking the survey. When the current round of data collection has been completed over 20 countries will have participated in the study worldwide.

The British Survey was carried out by the Social Survey Division of the Office for National Statistics. The survey was commissioned by a consortium of Government Departments, the main funding being provided by the Department for Education and Employment. Other contributing departments included the Department for Trade and Industry, the Scottish Office, the Department for Social Security, Socio-Economic Statistics and Analysis Group of the Office for National Statistics as well as the Basic Skills Agency (BSA). The survey was also carried out in Northern Ireland where it was carried out by the Central Survey Unit of the Northern Ireland Statistics and Research Agency.[2]

This survey is the first literacy survey to be carried out in Britain on a national random probability sample of adults of working age.[3] It set out to profile the literacy abilities of adults aged 16-65 using an internationally agreed measurement instrument and survey implementation protocols which covered among others, interviewer instructions and scoring procedures. This report presents some of the results of most immediate interest from the British survey. Chapter 6 makes comparisons with those countries that have already published their results. A second international report is due for publication by the Organisation for Economic Co-operation and Development (OECD) and Statistics Canada later in 1997 which will report on results from the expanded pool of countries.

1.2 What is literacy?

The International Adult Literacy Survey had its genesis in work carried out in the US during the 1980s, in particular the Young Adult Assessment (YAL).[4] This used open-ended rather than multiple choice assessment tasks as it more closely reflected the situation in which adults have to perform literacy tasks. The assessment tasks were taken from a broad range of contexts simulating the range of tasks that adults would encounter in everyday life. Unlike previous studies the YAL study did not treat literacy as a single dimension. Rather it identified three scales that represented three different aspects of literacy - prose, document and quantitative - which better reflected the diversity of literacy tasks that are encountered in daily life. The definition of literacy adopted for the international survey is that used in the YAL survey. It defines literacy as

Using printed and written information to function in society, to achieve one's goals and to develop one's knowledge and potential.

This definition does not treat literacy as a dichotomous condition that people either have or do not have, but rather defines literacy as a broad range of skills required in a varied range of contexts. It also implies that literacy goes beyond merely reading or comprehending text to include a broader range of skills in using information in texts. In IALS three dimensions of literacy skill are measured:

Prose literacy: the knowledge and skills required to understand and use information from texts such as prose, newspaper articles and passages of fiction. The texts have a typical paragraph structure.

Document literacy: the knowledge and skills required to locate and use information contained in various formats such as timetables, graphs, charts and forms. The texts have a varied format, use abbreviated and/or informal language and use a variety of devices and visual aids to convey meaning such as diagrams, maps or schematics.

Quantitative literacy: the knowledge and skills required to apply arithmetic operations, either alone or sequentially, to numbers embedded in printed materials, such as calculating savings from a sale advertisement, working out the interest required to achieve a desired return on an investment or totaling a bank deposit slip.

Each of the three scales which measure these dimensions of literacy skill is designed to range from 0 to 500 and has been grouped into five literacy levels. Level 1 represents the lowest ability range and Level 5 the highest. Each level, as shown on Figure 1.1, implies an ability to cope with a particular type of task and are based on incremental shifts in the skills required to successfully complete items at different points along the scales. Although the three scales are highly correlated individuals do not necessarily perform equally well on each scale.

All the stimuli included in the assessment were real items drawn from the countries taking part in the first round of the international survey. The items in the assessment represent a broad range of contexts and reflect the diversity, reality and challenges of everyday life.[5] The difficulty of an item is associated with the characteristics of the task and the attributes of the text so that performance on any particular task reflects the interaction between the characteristics of the task itself and both the context and the format of the text.

The IALS made use of Item Response Theory (IRT), a statistical method for scaling test items for difficulty so that the item has a known probability of being correctly completed by an individual with a given proficiency level. To be placed at a particular level on a scale respondents have to consistently perform tasks at that level correctly. The definition of consistent performance for the survey was set at 80%.[6] Individuals at Level 3 for example should perform

tasks at that level consistently - getting them right 80% of the time. They would have a higher than 80% probability of correctly answering lower level items. Similarly, they would sometimes be able to answer a higher level task correctly but they would not be able to perform items at higher levels consistently, that is, getting them right at least 80% of the time. Respondents received a score based on their performance on the literacy assessment. Some respondents only completed part of the assessment and where they had completed insufficient tasks to calculate their performance an imputation process was used to estimate their proficiency.

1.3 How IALS levels relate to other standards

The IALS framework for defining literacy levels is only one of many such classifications. Not only do different typologies identify different core or basic skills but they also define attainment in those skills in different ways. As a result there is no consensus on what skills comprise core/basic or key skills and no easy way to map one classification system onto another.[7] Many such classifications use models based to some extent on the interaction between the complexity of the material and the independence of the person carrying out the task. The Basic Skills Agency standards indentify four main skills: reading, writing, numeracy and oral communication. Within each skill area it differentiates between different aspects of skill. Reading skill is separated into reading texts, reading graphical materials and reading reference systems such as filing systems. Numeracy skills are broken down into four different types of tasks. Within each of these skill areas four levels of skill attainment are identified, Foundation level and Levels 1 to 3.

The National Council for Vocational Qualifications (NCVQ) classification of key skills, in addition to communication and application of numbers, also includes such skills as problem solving, team working and the ability to learn. Four levels of accomplishment are identified, Entry level, Foundation (NVQ1),

Figure 1.1 Description of the prose, document and quantitative literacy levels

Level	Prose	Document	Quantitative
Level 1 (0-225)	Locate one piece of information in a text that is identical or synonymous to the information in the question. Any plausible incorrect answer present in the text is not near the correct information	Locate one piece of information in a text that is identical to the information in the question. Distracting information is usually located away from the correct answer. Some tasks may require entering given personal information on a form.	Perform a single simple operation such as addition for which the problem is already clearly stated or the numbers are provided
Level 2 (226-275)	Locate one or more pieces of information in a text but several plausible distractors may be present or low level inferences may be required. The reader may also be required to integrate two or more pieces of information or to compare and contrast information.	Tasks at this level are more varied. Where a single match is required more distracting information may be present or a low level inference may be required. Some tasks may require information to be entered on a form or to cycle through information in a document.	Single arithmetic operation (addition or subtraction) using numbers that are easily located in the text. The operation to be performed may be easily inferred from the wording of the question or the format of the material.
Level 3 (276-325)	Readers are required to match information that require low-level inferences or that meet specific conditions. There may be several pieces of information to be identified located in different parts of the text. Readers may also be required to integrate or to compare and contrast information across paragraphs or sections of text.	Literal or synonymous matches in a wide variety of tasks requiring the reader to take conditional information into account or to match on multiple features of information. The reader must integrate information from one or more displays of information or cycle through a document to provide multiple answers.	At this level the operations become more varied - multiplication and division. Sometimes two or more numbers are needed to solve the problem and the numbers are often embedded in more complex texts or documents. Some tasks require higher order inferences to define the task.
Level 4 (326-375)	Match multiple features or provide several responses where the requested information must be identified through text based inferences. Reader may be required to contrast or integrate pieces of information sometimes from lengthy texts. Texts usually contain more distracting information and the information requested is more abstract.	Match on multiple features of information, cycle through documents and integrate information. Tasks often require higher order inferences to get correct answer. Sometimes, conditional information in the document must be taken into account in arriving at the correct answer.	A single arithmetic operation where the statement of the task is not easily defined. The directive does not provide a semantic relation term to help the reader define the task.
Level 5 (376-500)	Locate information in dense text that contain a number of plausible answers. Sometimes high-level inferences are required and some text may use specialized language.	Readers are required to search though complex displays of information that contain multiple distractors, to make high-level inferences, process conditional information or use specialised language.	Readers must perform multiple operations sequentially and must state the problem from the material provided or use background knowledge to work out the problem or operations needed.

Intermediate (NVQ2) and Advanced (NVQ3). These correspond, broadly to the BSA levels, Foundation (or Entry) Level and Levels 1 to 3.

Establishing equivalence between the different classifications is not easy. Not only do they define basic skills in different ways but they also have different underlying conceptual frameworks and varying thresholds for defining competency. One important difference, for example, between the Basic Skills Agency standards and the IALS levels is that the IALS set out to measure the full range of ability whereas the BSA standards do not extend to the highest literacy levels. Without undertaking a detailed analysis of tasks representative of each level for the different standards it is not possible to establish the congruence of the different hierarchies. However, at the two extremes it is possible to say that part of IALS Level 4 and all of IALS Level 5 represent a higher order of skills than the tasks at the highest BSA standard (Level 3) and that the tasks at BSA Foundation (or Entry) level and some of the tasks at BSA Level 1 for reading fall within the IALS Level 1.

1.4 Making international comparisons

One of the objectives of IALS was to be able to examine the determinants of literacy across a number of countries, languages and cultures. In order to meet this objective all countries taking part in the survey were required to implement a common survey design using common instruments. The specification stipulated a number of key design features which included:

- a minimum sample size and response rate
- the use of probability sampling
- quality checking the scoring of items by rescoring 20% of assessments within the country and 10% by another country where language permitted
- the use of the common instruments (background questionnaire and literacy assessment) which were developed collaboratively by countries taking part in the first round
- asking all mandatory questions on the background

questionnaire with minimum modification
- administering the literacy assessment in exactly the format prescribed (for example, font, layout, size, structure) with modifications limited to adaptations of text to local cultural context and usage
- coding industry, occupation and education to standard international classifications
- implementing the survey in a standard way, for example, allowing the respondent as much time as they required, not using incentives and using personal interview for the background questionnaire
- weighting the data to known population estimates and to take account of non-response

Despite every effort to ensure uniformity not all countries met the standardised design specification and there are differences between the countries in how the surveys were implemented. These issues are not by any means unique to IALS but apply to all international surveys. In particular there were differences in the response rates achieved, the sampling frames and methods used and the data collection procedures. For example, response rates varied from 45% in the Netherlands to 69% in Germany. However, the comparability of estimates from samples in different countries is affected by the presence or absence of non-response bias which is not necessarily indicated by low response to the survey. This is quite apart from any more general concerns about question translation, meaning or appropriateness in the different countries or languages.[8] A description of the survey design and response for each participating country will be published in the IALS Technical Report.[9] Although the effect of some of these differences is unknown and for others it is expected to be small they mainly affect comparisons of the overall distributions of literacy skills between countries, ie, across different samples. Analysis of relationships within countries, which was one of the main objectives of the survey will not be affected. It is recommended as a result, that comparisons of results across different countries should be undertaken with due caution.[10] As IALS was conceived as an international survey, in many aspects, such as common instruments and assessment administration, the survey offers greater comparability than some other international data sets.

1.5 How the survey was conducted

The survey was conducted by personal interview in respondents' homes and consisted of two main elements, a background questionnaire and a literacy assessment. Both instruments were developed collaboratively by countries participating in the first round. The background questionnaire collected information on the socio-demographic characteristics of the respondent such as age, sex, education, occupation and income as well as asking about literacy activities such as reading as part of their job or for pleasure, television viewing, participation in training or adult education. For the purpose of making international comparisons the background questionnaire had certain questions designated as core questions which every country had to include. Where the international questionnaire did not fit a local situation a certain amount of modification could be made to make the questionnaire appropriate and relevant to local conditions. In Britain the background questionnaire and assessment administration used Computer Assisted Interviewing methods.

After the background questionnaire, respondents were asked to complete a short screening assessment which sought to identify those with very limited literacy skills. Respondents who correctly answered at least 2 of the 6 screening tasks were then presented with a larger assessment booklet to complete which measured the three dimensions of literacy. Although respondents had to write their answers in the booklet, the assessment did not measure writing ability. In order to ensure as broad a range of item content as possible the total number of tasks in the assessment was larger than any one individual could reasonably be asked to complete. Each respondent therefore was only asked to complete a subset of the total assessment. The assessment items were grouped into seven blocks and a Balanced Incomplete Block (BIB) design was used to arrange the blocks in different

combinations into seven booklets. Each booklet contained 3 blocks of items and each block appeared at each possible location, the beginning, middle or end of a booklet in a spiral effect. Respondents were allowed to take as much time as they required to complete the booklet. Some sample items from the assessment are shown in Appendix D.

The booklets were scored using a scoring guide common to all countries. The scoring differentiated between correct answers, incorrect answers and items which had not been attempted. Sequences of consecutive items coded as not attempted were used to identify incomplete assessments.

1.6 Limitations of sample size

Most countries had modest sample sizes and this is true of Britain as well. The overall achieved sample in Britain was 3,811, 68% of eligible respondents. This limits the extent to which the proficiency estimates of some sub-populations can be reported. In the British sample Scotland and Wales were over-represented in order to be able to provide separate ability estimates for both but the sample sizes are nevertheless limited in the analysis they will support. For many variables categories need to be combined before presenting results by country. The survey design and response for Great Britain is described in Appendix A.

Notes and references

1 *Literacy, Economy and Society: Results of the first International Adult Literacy Survey.* OECD and Statistics Canada (1995).

2 The Northern Ireland sample was equivalent to that for some countries in order to support more detailed analysis than if it was simply proportional to a UK sample.

3 Previous literacy surveys carried out in Britain were based on cohort studies or did not use probability sampling at all stages of the sample design so that each respondent did not have an equal probability of selection. Probability sampling, where each respondent has an equal probability of selection, allows the use of statistical technique to generalise to the population.

4 Irwin S Kirsch and Ann Jungeblut. *Literacy: Profiles of America's Young Adults.* Educational Testing Service (1986 Princeton, N.J).

5   For a fuller description of the assessment methodology see Irwin Kirsch. Literacy performance on three scales: definitions and results. In *Literacy, Economy and Society* OECD and Statistics Canada (1995) pp27-54.

6   There has been some discussion about the appropriateness of the 80% criterion for similar studies. The effect of altering the criterion on the National Adult Literacy Survey (NALS) from 80% to 65% or 50% resulted in reducing the proportion of adults below level 3 from 47% to 32% and 22% respectively. See Thomas G. Sticht. How many marginally literate adults are there? In the *U.S. Journal of Educational Research* Research Note 20 (April 1996).

7   See for example discussion in Peter Robinson. *Literacy and Numeracy and Economic Performance.* Working Paper No. 888, Centre for Economic Performance, London School of Economics (June 1997).

8   Social Survey Division is currently carrying out a study to investigate cultural bias in comparative survey research using the IALS as the basis for the research.

9   *IALS Technical Report.* National Center for Educational Statistics (forthcoming).

10 For a further discussion of problems in making international comparisons on studies using a similar methodological framework to IALS see Harvey Goldstein. Interpreting international comparisons of student achievement. *Educational studies and documents 6.* UNESCO (1995, Paris) and David Reynolds and Shaun Farrell. *Worlds Apart? A Review of International Surveys of Educational Achievement involving England.* OFSTED Reviews of Research. OFSTED (1996, London).

2 Distribution of literacy skills

2.1 Introduction

The survey measured three dimensions of literacy skill, prose, document and quantitative literacy. Each dimension was measured on a scale which ranges from 0 to 500 and which has been grouped into five literacy levels, each level representing the ability to cope with a particular type of task and based on incremental shifts in the skills required to successfully complete tasks at different points along the scales. Level 1 represents the lowest ability range and Level 5 the highest. Although the three scales represent three different aspects of literacy, as one might expect given the description of the three skill dimensions in Chapter 1, they are highly correlated (greater than 0.9).

As discussed in Chapter 1, the measure of literacy used in this survey has moved away from the traditional dichotomous definition to one based on a continuum. Defining the scales in terms of levels tempts the creation of a new dichotomy, however, there is no generally agreed interpretation on the minimum skill threshold for functioning in modern society. While it is clear that those at Level 1 demonstrate a poor ability to deal with a wide range of tasks the implications of being at Level 2 or Level 3 are not so easy to determine. Without understanding more about the literacy demands of society it is unhelpful to set arbitrary thresholds.

This chapter describes the distribution of these three dimensions of literacy skill for the population of working age (16-65) living in private households in Great Britain. Because the levels do not represent equal intervals on the scales, for the most part results are reported for the different levels of literacy rather than average literacy proficiency scores, although mean scores are included in many of the tables. Because of the small proportion of people at the highest level, Level 5, data are presented for Levels 4 and 5 combined (Level 4/5).

In reporting the distributions of literacy there are two main aspects of interest. First is the distribution of literacy skills in the population for key characteristics such as age, educational attainment level and economic activity status. Second is the extent to which there is internal consistency between the three dimensions of literacy when controlling for these characteristics.

2.2 Literacy skills of adults of working age

The distribution of literacy skills among the British population of working age is broadly similar for each of the three dimensions, prose, document and quantitative. Figure 2.1 shows the proportions of the population of working age at each literacy level for the three scales. Slightly less than a third (31%) of the population aged 16-65 performed at Level 3 on each of the three dimensions with the proportions at the other literacy levels varying slightly between the three scales. On the prose scale, 30% performed at Level 2 with slightly lower proportions performing at that level on document and quantitative literacy. Although the majority perform at Levels 2 and 3, considerable proportions of the population aged 16-65 perform at the lower and upper literacy levels, Level 1 and Level 4/5, on all three dimensions. Overall just over a fifth of the population of working age performed at Level 1 on the prose scale while 17% performed at the two highest levels combined (Level 4/5). Similarly, 23% of the adult population aged 16-65 performed at Level 1 on the document and quantitative scales. The performance of Britain relative to the other countries that took part in the study is discussed in Chapter 6.

(Figure 2.1, Table A2.1)

The following sections show the distribution of literacy for some of the main socio-demographic characteristics of the population. There is considerable variation in the literacy ability of the population across such characteristics as age, educational attainment level and economic activity status. Many of these variables are inter-related, for example, age and educational attainment level and so any apparent association (either positive or negative) between literacy skills and a particular characteristic may simply be due to the underlying association

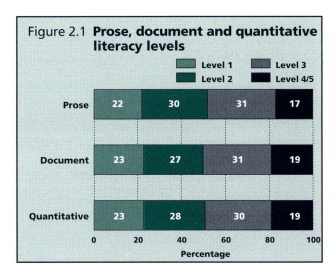

Figure 2.1 **Prose, document and quantitative literacy levels**

	Level 1	Level 2	Level 3	Level 4/5
Prose	22	30	31	17
Document	23	27	31	19
Quantitative	23	28	30	19

Percentage

between that characteristic and some other variable. In order to control for these effects multi-variate statistical techniques were used to identify those variables that were independently associated with literacy after controlling for the other variables entered in the model. Initially the models were run including only socio-demographic characteristics. Behavioural factors such as reading practice were then added in to the model to determine whether they contributed to explaining the variation in literacy skills. Modelling is used to identify the characteristics that have the greatest impact on literacy with the most important characteristic being entered first, which as might be expected in the case of literacy was educational attainment level.

(Table A2.11)

The characteristics which contributed most to explaining variance in prose literacy in order of importance were

- highest level of educational attainment
- social class
- age
- whether the respondent reads a book at least once a week
- whether English was the language first spoken as a child
- whether the respondent watches 5 or more hours television daily
- whether the respondent is in receipt of state benefits
- personal income from all sources
- ethnic group.

In general the order of importance was similar on all three scales but personal income from all sources was a more important factor in contributing to document and quantitative literacy level than it was for prose. Sex was also an important factor on the document and quantitative scales.

The remaining sections of this chapter describe the interactions of literacy and some of the socio-demographic characteristics found to be significantly associated with literacy. The effect of behavioural practices on literacy are discussed in Chapter 4, Literacy skills in everyday life.

2.3 **Literacy skills and gender**

For both men and women performance varied across the three literacy dimensions. Among men, performance on the prose scale was poorer than on either the document or quantitative scales whereas women performed better on prose than on the other two dimensions. About a quarter of men performed at the highest literacy level (Level 4/5) on both document and quantitative literacy compared with 17% who performed at that level on prose. Although there was no significant difference in the percentage of women performing at Level 1 on prose and document, (22% and 27% respectively), a significantly higher proportion of women performed at Level 1 on quantitative literacy (29%) than on prose.

While the skill levels of men and women were very similar on the prose scale the difference in performance over the three literacy dimensions for men and women operates in different directions so that significantly higher proportions of women than men were at Level 1 on both document and quantitative literacy. On document literacy 27% of women performed at Level 1 compared with 20% of men. On the quantitative scale the differences between the sexes was even greater with 29% of women at Level 1 compared with 18% of men.

(Figure 2.2, Table A2.1)

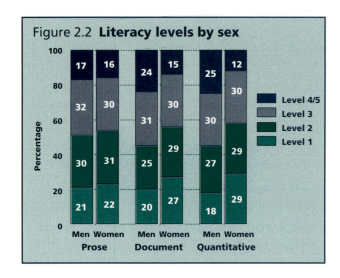

Figure 2.2 **Literacy levels by sex**

age-groups move out of the available labour market the overall distribution of literacy skills in the population of working age should improve assuming that the lower performance of the older age-groups is attributed to lower educational levels and that there is no skill loss among those in the lower age ranges with advancing years.[1] The corollary, however, does not appear to be true in that there is no apparent improvement to the overall distribution of skills to be gained from the youngest age-group. There is a plateau effect among the three youngest age-groups which had very similar distributions so that unless proficiency in the skills measured increase through practical application of these skills in the workplace among new labour market entrants there is no prospect of substantial improvement in the overall skill profile of the labour force except through the exit of the older age cohorts.

(Tables 2.1, A2.1)

2.4 **Literacy skills and age**

Table 2.1 shows the distribution of prose, document and quantitative literacy skills by age-group. On all three scales there was a higher proportion performing at Level 1 in the two oldest age-groups, that is, those aged over 45 than in the younger age-groups. Because of changes in the participation rates in education over the years younger people are more likely to have higher levels of educational attainment than older age-groups and therefore might be expected to have higher literacy levels. As the older

All estimates based on sample surveys are subject to error, some of which are measurable. One measure of the reliability of a survey estimate is the standard error which allows us to estimate how close the results from this survey are likely to be to the true population value. The precision of a survey estimate is related to

Table 2.1	**Literacy level by age-group**										
	Level 1		Level 2		Level 3		Level 4/5		Total	Mean score	*Base*
	%	s.e.	%	s.e.	%	s.e.	%	s.e.	%		
Prose literacy											
16-25	17	1.7	30	2.4	33	2.7	20	1.8	100	274	*549*
26-35	18	1.5	29	2.0	34	2.0	19	1.5	100	275	*991*
36-45	17	1.3	29	2.3	33	2.2	21	1.6	100	277	*844*
46-55	22	2.6	29	2.7	35	2.4	14	1.3	100	264	*724*
56-65	39	2.2	37	2.2	19	1.5	6	1.2	100	236	*703*
Total	22	1.0	30	1.3	31	1.2	17	0.8	100	267	*3811*
Document literacy											
16-25	18	1.8	27	1.9	34	2.4	22	2.0	100	276	*549*
26-35	19	1.6	25	2.2	32	1.8	23	1.6	100	278	*991*
36-45	19	1.7	24	2.3	32	2.0	24	1.8	100	278	*844*
46-55	24	2.3	28	1.7	31	2.9	16	1.6	100	264	*724*
56-65	40	2.4	33	2.3	21	1.3.	6	1.0	100	233	*703*
Total	23	1.0	27	1.0	31	1.0	19	1.0	100	268	*3811*
Quantitative literacy											
16-25	22	1.7	29	2.5	33	2.4	16	2.1	100	265	*549*
26-35	20	1.7	28	2.0	30	2.0	23	1.6	100	277	*991*
36-45	19	1.2	24	2.0	32	1.9	25	1.5	100	279	*844*
46-55	24	2.3	26	2.0	33	2.8	17	1.7	100	266	*724*
56-65	35	2.3	34	2.5	23	1.8	8	0.9	100	240	*703*
Total	23	0.9	28	1.0	30	0.9	19	1.0	100	267	*3811*

s.e. = Standard error of the estimate. The reported sample estimate can be said to be within 2 standard errors of the true population value with 95% confidence the variability of the characteristic of interest, in this case literacy skill, the size of the sample and the sample design. The larger the sample the smaller the standard error will be and the smaller the standard error the more confident we can be of the estimates reliability. As the size of the subgroup decreases the standard error increases making the estimate less reliable.

To illustrate the importance of taking account of sampling error, Figure 2.3 shows the 95% confidence intervals around the estimate of those at Level 1 on the quantitative scale for each age-group. If one hundred separate samples were taken, the estimate of those at Level 1 in each age group would fall within the ranges shown in 95 of those samples. In terms of statistical significance, where there is an overlap in the confidence intervals the differences between the estimates for those groups will not be significant. So for example, although a higher proportion of the very youngest age-group performed at Level 1 on quantitative literacy compared with the next two age-groups, we can see that this difference was not statistically significant as the range of estimates overlap.

(Figure 2.3, Table A2.1)

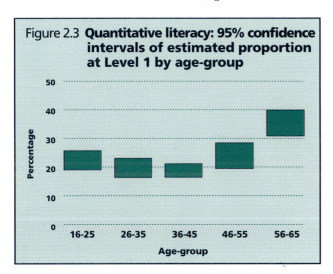

Figure 2.3 **Quantitative literacy: 95% confidence intervals of estimated proportion at Level 1 by age-group**

2.5 Classification of educational attainment

There are different ways of measuring educational attainment. One method takes as its focus the completion of educational programmes at different levels while another is based on the attainment of qualifications. To enable international comparisons to be carried out each country was required to code highest level of education completed to the International Standard Classification of Education (ISCED).[2] ISCED divides educational attainment into 7 categories which span three broad levels of education roughly equivalent to primary, secondary and tertiary education.

ISCED 0   Education preceding the first level, if provided, usually begins at age 3, 4 or 5 and lasts one to three years (pre-primary)

ISCED 1   First level education, usually begins at age 5, 6 or 7 and lasts for about five or six years (primary)

ISCED 2   Second level, first stage begins at about age 11 and lasts for about 3 years (lower secondary)

ISCED 3   Second level second stage education begins at about age 14 or 15 and lasts for about three years (higher or upper secondary)

ISCED 4   Not used

ISCED 5   Third level or higher education which leads to an award which is not equivalent to a university degree, for example a Higher National Diploma (HND)

ISCED 6   Third level or higher education that leads to a university degree or equivalent

ISCED 7   Third level or higher education post first degree that leads to a post-graduate university degree or equivalent.

ISCED 9   Education not definable by level

The ISCED is commonly used to compare educational outputs and the relative performance of one education system against another. However, the ISCED is currently being revised as it is no longer felt to adequately capture the complexity and structural diversity of modern education and qualification systems. One of the anomalies of ISCED is that very similar courses are often classified at different ISCED levels in different countries. This can affect the overall distribution of ISCED quite considerably and limits the interpretation of international comparative education statistics.

The ISCED classifications are very broad and each country has to fit their own educational framework on to the classification. There are discrepancies in how different international organisations map the UK educational system onto the ISCED codes. This centres around where the boundary between ISCED 2 (Second level, first stage) ends and ISCED 3 (Second level, second stage) begins, specifically where people whose highest qualifications are GCSEs or equivalent are classified. On one version of ISCED (that used by EUROSTAT) these people are classified at lower secondary education (ISCED level 2) whereas on the other version (used by the OECD) the same

people are classified as upper secondary education (ISCED level 3). As a large proportion of the population have GCSEs or equivalent as their highest qualification, where this boundary is drawn greatly affects the distribution of educational attainment levels in Britain. When using one classification the majority of the British population would appear to be qualified up to ISCED level 2 whereas the other classification has the majority of Britons qualified up to ISCED level 3. The ISCED version which shows the majority of the population at ISCED level 2 was used for this study. [3]

2.6 Literacy skills and educational attainment

For IALS, countries were required to code education to the highest level of education completed. However, as this is not a standard classification used in Britain,

Table 2.2 Literacy level by level of highest qualification

	Level 1 %	Level 2 %	Level 3 %	Level 4/5 %	Total %	Mean score	Base
Prose literacy							
Degree or equivalent	3	13	39	46	100	316	539
Other Higher Education below degree level	4	21	49	26	100	302	415
A-Levels, vocational level 3 & equivalents	5	25	43	26	100	296	454
Trade apprenticeships	26	42	27	6	100	253	335
GCSE/O Level grade A*-C, vocational level 2 & equivalents	9	31	40	20	100	285	668
Qualifications below level 2	19	48	27	6	100	260	275
Other qualifications - level unknown	47	32	18	3	100	223	135
No qualifications	48	33	16	3	100	221	990
Total	**22**	**30**	**31**	**17**	**100**	**267**	**3811**
Document literacy							
Degree or equivalent	4	12	35	50	100	320	539
Other Higher Education below degree level	7	20	45	28	100	301	415
A-Levels, vocational level 3 & equivalents	7	20	41	32	100	300	454
Trade apprenticeships	30	33	30	8	100	255	335
GCSE/O Level grade A*-C, vocational level 2 & equivalents	10	28	39	23	100	288	668
Qualifications below level 2	18	43	30	9	100	262	275
Other qualifications - level unknown	49	23	25	3	100	221	135
No qualifications	49	32	15	4	100	218	990
Total	**23**	**27**	**31**	**19**	**100**	**268**	**3811**
Quantitative literacy							
Degree or equivalent	3	12	29	56	100	325	539
Other Higher Education below degree level	7	22	42	30	100	300	415
A-Levels, vocational level 3 & equivalents	7	23	41	30	100	298	454
Trade apprenticeships	26	35	31	8	100	257	335
GCSE/O Level grade A*-C, vocational level 2 & equivalents	13	27	41	19	100	283	668
Qualifications below level 2	25	38	29	8	100	259	275
Other qualifications - level unknown	46	27	25	3	100	224	135
No qualifications	47	34	16	3	100	219	990
Total	**23**	**28**	**30**	**19**	**100**	**267**	**3811**

and because of the different ways ISCED can be applied, in this chapter, results are presented using both highest level of qualification and highest level of education completed.

2.6.1 Literacy skills and highest level of qualification

One measure of educational achievement is the highest level of qualification held. As might be expected performance on the literacy scales was strongly associated with educational qualifications, the percentage of people performing at the higher literacy levels increasing with increasing education. On each of the three scales, almost half of those with no qualifications were at Level 1. Among those with GCSE, vocational level 2 or equivalent qualifications, on prose the majority performed at the middle literacy levels, 40% at Level 3 and 31% at Level 2 and a further 20% were at the highest literacy level, Level 4/5. The distribution of skills among those with A level, vocational level 3 or equivalent and those with higher education below degree level were very similar. On the prose scale, just over a quarter (26%) were at Level 4/5 and large proportions were at Level 3. Almost half (49%) of those with higher education below degree level were at Level 3 on prose as were 43% of those with A level or equivalent qualifications. Among those with degree or equivalent qualifications almost half (46%) were at prose Level 4/5 and 39% were at Level 3.

(Table 2.2, Table A2.2)

2.6.2 Literacy skills and highest level of education completed

Figure 2.4 shows the distribution of literacy skills using the ISCED. As would be expected it shows the same pattern observed above. Those with lower levels of education were much more likely to be at literacy Levels 1 and 2. On each of the three scales, half of those with primary or lower education were at Level 1. Of those with university education (ISCED levels 6 and 7) almost half (46%) were at Level 4/5 on prose with even higher proportions performing at that level on document and quantitative literacy (50% and 56%) respectively.

Although literacy performance is strongly associated with education whichever classification is used there are a minority of people for whom it is not a good indicator of their literacy skill. A small percentage at both ends of the education spectrum did not fit the expected pattern, (about 5% in total). On each of the three scales, among those with third level education or qualifications a small proportion of individuals performed at Level 1 and around one fifth performed at the two lowest literacy levels combined. Literacy is not a skill that is acquired solely in the context of formal education as some of those with minimal schooling do perform at very high proficiency levels. A small percentage of those with primary education or lower performed at Level 4/5. Equally, assuming that those with relatively high levels of education did at one time have higher literacy proficiencies, it would suggest that individuals can lose skills after leaving the education system perhaps through lack of use or practice.

(Figure 2.4, Table A2.3)

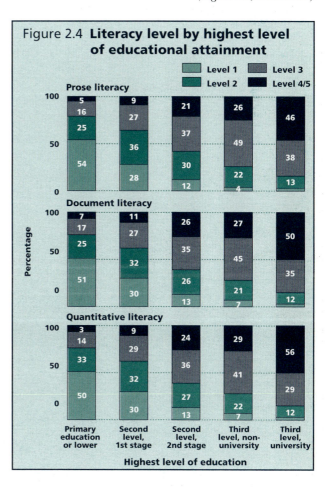

Figure 2.4 **Literacy level by highest level of educational attainment**

2.7 Literacy skills and economic activity status

There were considerable differences between the literacy proficiencies of those in employment or in full-time education and those who were unemployed or economically inactive. Those in employment and full-time students were more likely to perform at Level 4/5 than the unemployed or economically inactive on all three scales. A fifth of the employed were at Level 4/5 on the prose scale with slightly higher proportions performing at that level on the document and quantitative scales.

The unemployed were twice as likely as those in employment to perform at Level 1. On each of the three scales, almost a third of the unemployed were at Level 1 compared with 16% of the employed. Around two-thirds of the unemployed were at the two lowest literacy levels combined. Among the retired and economically inactive very large proportions performed at the lower literacy levels on all three scales. Part of this is explained by the underlying association between employment status and educational attainment level. Those with no qualifications are less likely to be employed than those with A level or higher qualifications. Education however is also associated with age. When educational attainment level was held constant, among those qualified to A level, vocational level 3 or higher there was no significant difference in the literacy profiles of the employed and unemployed. For those whose highest level of qualification is GCSE, vocational level

2 or equivalent the unemployed were more likely to perform at Level 1 than the employed. The same was true for those with no qualifications.

(Figure 2.5, Tables A2.4 and A2.5)

2.8 Literacy skills and social class as defined by occupation

In order to code social class, informants were asked to give details of their current or most recent job, or if retired, of their main job during their working life. This information was coded to six social class groups using the Standard Occupational Classification.[4] As the social class of the respondent is used rather than social class of head of household, for some respondents, particularly full-time students and married women, their current or most recent jobs may not be a very good guide to their economic and social status. Informants who were members of the Armed Forces or whose occupation has been inadequately described or who never worked were not allocated a social class. Social class has been presented as four categories:

I and II	Professional, managerial and technical occupation
III (non-manual)	Skilled non-manual occupations
III (manual)	Skilled manual occupations
IV and V	Unskilled occupations.

There was a clear distinction between the performance of those in manual social class groups compared with those in non-manual groups. On all three literacy scales those in manual social class groups were much more likely than others to perform at the lower literacy levels. Over a third (36%) of those in Social Classes IV and V and 29% of those in Social Classes III (manual) were at Level 1 on prose compared with 12% or less for other social class groups. The manual social class groups had very similar mean proficiency scores on prose (242 and 248) as did the non-manual social groups (283 and 294).

Although the percentage of those in Social Classes I and II performing at Level 1 was the same on each of

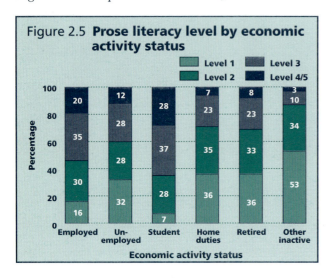

Figure 2.5 **Prose literacy level by economic activity status**

the three scales, the percentage performing at Levels 4/5 was higher for document and quantitative literacy than for prose. On quantitative literacy 35% of those in the Social Classes I and II performed at Level 4/5 while 28% performed that high on prose. The distribution of literacy skills according to occupation is shown in Chapter 3.

(Figure 2.6, Table A2.6)

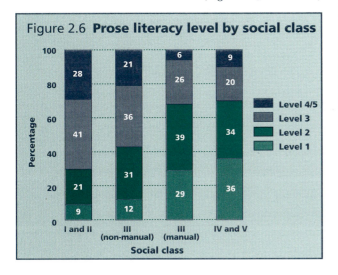

Figure 2.6 **Prose literacy level by social class**

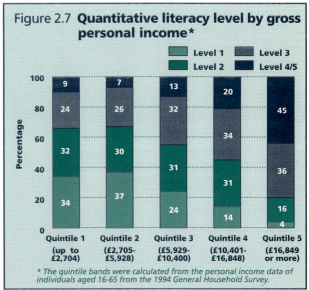

Figure 2.7 **Quantitative literacy level by gross personal income***

* The quintile bands were calculated from the personal income data of individuals aged 16-65 from the 1994 General Household Survey.

2.10 Literacy skills and receipt of benefits

As part of the income section respondents were asked which social security benefits, if any, they were receiving. Those in receipt of social security benefits (excluding pensions and child benefit) were much more likely to have low literacy skills than those not in receipt of social security benefits. Of those who were receiving such benefits, 40% were at Level 1 on prose compared with 17% of those not receiving benefit. On document and quantitative 44% and 45% of those receiving benefits were at Level 1 with only 8% and 5% of those performing at Level 4/5. Among those whose sole source of income was social security benefits over two thirds were at literacy Levels 1 or 2 on the document scale, with slightly less than half performing at Level 1. (Table not shown).

(Table 2.3)

2.9 Literacy skills and gross income

The income data that was collected as part of the interview was not very detailed. Banded income quintiles were used to facilitate international comparisons by avoiding problems of establishing equivalence across the different countries. Quintiles divide the distribution into five equal parts so that each quintile covers 20% of the population.[5] On all three literacy scales, those in the lowest income quintile were much more likely to perform at Level 1 than those in the two highest income quintiles. Thirty percent of those in the lowest income range and 32% of those in the second lowest income quintile performed at Level 1 on prose with even higher proportions performing at that level on document and quantitative literacy. Only 12% of those in the lowest income quintile performed at Level 4/5 on prose. By contrast almost a third of those in the highest income quintile performed at the highest literacy levels combined (Level 4/5) on prose and 44% performed at that level on the quantitative scale.

(Figure 2.7, Table A2.7)

Table 2.3   Literacy level by whether or not receives social security benefits (excluding pensions and child benefit)

	Level 1	Level 2	Level 3	Level 4/5	Total	Mean score	Base
Prose literacy	%	%	%	%	%		
No	17	30	34	19	100	276	2908
Yes	40	31	21	7	100	231	876
Total	22	30	31	17	100	267	3784
Document literacy							
No	18	27	33	22	100	277	2908
Yes	44	28	20	8	100	226	876
Total	23	27	31	19	100	268	3784
Quantitative literacy							
No	18	28	32	22	100	277	2908
Yes	45	27	23	5	100	226	876
Total	23	28	30	19	100	267	3784

2.11 Literacy skills, country of birth and language first spoken as a child

Only a small percentage of the sample (7%) were born outside of the United Kingdom. On the face of it, their literacy proficiencies differ considerably from those born in the UK with 41% of non-UK born at Level 1 on prose compared with 20% of UK born. Although the difference is statistically significant, because it is based on a small sub-population the standard errors of the estimates for non-UK born are high (4.7 s.e. for the Level 1 estimate). When we take into account the standard error for the estimate of non-UK born performing at Level 1 on prose, we can say that in 95 out of 100 samples the estimate of those born outside the UK who are at Level 1 on the prose scale would be between 32% and 50%.

(Table 2.4)

Again only a small proportion of the sample (6%) reported that the language they had first spoken as a child was not English. Respondents who first spoke a language other than English were more than twice as likely to perform at the lower end of the proficiency range than those who first spoke English. Only 5% of those who first spoke a language other than English were at Level 4/5 on the prose scale compared with 17% of those who learned English as their first language. As with other estimates based on small sub-populations, this estimate is rather imprecise. The 95% confidence interval for the estimate of those who first spoke a language other than English who are at Level 1 is between 36% and 58%. As the purpose of the study was to estimate literacy proficiency in English, those whose first language is a language other English may well have higher literacy abilities in their first language.

(Table 2.5)

Table 2.4   Literacy level by whether born in the UK

	Level 1		Level 2		Level 3		Levels 4/5		Total	Mean score	Base
	%	s.e.	%	s.e.	%	s.e.	%	s.e.	%		
Prose literacy											
Born in the UK	20	0.8	31	1.3	32	1.2	17	0.8	100	270	3564
Born outside the UK	41	4.7	23	3.1	24	4.8	12	2.6	100	230	247
Total	22	1.0	30	1.3	31	1.2	17	0.8	100	267	3811
Document literacy											
Born in the UK	21	0.8	28	1.0	31	0.9	20	1.0	100	271	3564
Born outside the UK	44	4.7	21	3.1	20	4.0	15	3.0	100	230	247
Total	23	1.0	27	1.0	31	1.0	19	1.0	100	268	3811
Quantitative literacy											
Born in the UK	22	0.7	28	1.1	31	1.0	19	1.1	100	270	3564
Born outside the UK	41	5.0	26	3.2	18	3.7	15	2.7	100	235	247
Total	23	0.9	28	1.0	30	0.9	19	1.0	100	267	3811

s.e. = Standard error of the estimate. The reported sample estimate can be said to be within 2 standard errors of the true population value with 95% confidence

Table 2.5   Literacy level by whether English was the language first spoken as a child

	Level 1		Level 2		Level 3		Levels 4/5		Total	Mean score	Base
	%	s.e.	%	s.e.	%	s.e.	%	s.e.	%		
Prose literacy											
English	20	0.8	30	1.3	32	1.1	17	0.8	100	270	3557
Not English	47	5.3	32	4.1	17	4.6	5	1.5	100	213	254
Total	**22**	**1.0**	**30**	**1.3**	**31**	**1.2**	**17**	**0.8**	**100**	**267**	**3811**
Document literacy											
English	22	0.8	27	1.0	31	1.0	20	1.0	100	271	3557
Not English	48	5.6	24	3.3	18	4.0	10	2.5	100	215	254
Total	**23**	**1.0**	**27**	**1.0**	**31**	**1.0**	**19**	**1.0**	**100**	**268**	**3811**
Quantitative literacy											
English	21	0.7	28	1.1	31	0.9	19	1.1	100	270	3557
Not English	51	5.8	19	2.6	21	4.6	8	2.1	100	220	254
Total	**23**	**0.9**	**28**	**1.0**	**30**	**0.9**	**19**	**1.0**	**100**	**267**	**3811**

s.e. = Standard error of the estimate. The reported sample estimate can be said to be within 2 standard errors of the true population value with 95% confidence.

Table 2.6   Literacy levels in England, Scotland and Wales

	Level 1		Level 2		Level 3		Levels 4/5		Total	Mean score	Base
	%	s.e.	%	s.e.	%	s.e.	%	s.e.	%		
Prose literacy											
England	21	1.2	30	1.4	31	1.3	17	0.9	100	267	2472
Scotland	23	2.0	32	1.8	31	2.4	14	2.1	100	266	704
Wales	24	1.8	34	2.3	33	2.6	9	1.3	100	258	635
Total	**22**	**1.0**	**30**	**1.3**	**31**	**1.2**	**17**	**0.8**	**100**	**267**	**3811**
Document literacy											
England	23	1.2	26	1.1	31	1.2	20	1.2	100	268	2472
Scotland	22	2.1	31	1.9	29	2.5	17	2.3	100	267	704
Wales	26	1.6	31	2.2	30	2.0	13	1.8	100	258	635
Total	**23**	**1.0**	**27**	**1.0**	**31**	**1.0**	**19**	**1.0**	**100**	**268**	**3811**
Quantitative literacy											
England	23	1.1	27	1.2	30	1.1	19	1.2	100	268	2472
Scotland	24	2.2	30	1.7	31	2.3	15	1.0	100	266	704
Wales	25	1.5	31	2.2	32	2.5	12	1.7	100	260	635
Total	**23**	**0.9**	**28**	**1.0**	**30**	**0.9**	**19**	**1.0**	**100**	**267**	**3811**

s.e. = Standard error of the estimate. The reported sample estimate can be said to be within 2 standard errors of the true population value with 95% confidence.

2.12 Distribution of literacy skills in England, Scotland and Wales

The Adult Literacy Survey was carried out throughout Great Britain and the following section looks at the data separately for England, Scotland and Wales. Scotland and Wales were over-represented in the sample in order to be able to provide separate estimates for each. Nevertheless the sample sizes for Scotland and Wales limit the extent to which the data can support analysis. Of the total interviews, 2,472 were carried out in England, 704 in Scotland and 635 in Wales.

There was no difference in the mean proficiency scores or the literacy level distribution for England and Scotland on any of the three literacy scales. Wales had slightly lower mean scores (about 10 points lower) on all three scales. In terms of levels, on all three dimensions there were significant differences between England and Wales in the proportion of adults at Level 4/5 with no significant differences between England and Scotland in the proportion of adults who were at this level. The biggest difference was on prose with 17% of those living in England performing at Level 4/5 compared with 9% of those living in Wales.

(Tables 2.6 and A2.8)

29

Figure 2.8 shows the 95% confidence intervals around the estimates of the proportions of adults at Level 4/5 in each country for quantitative literacy. The confidence intervals illustrate that although there is a significant difference between England and Wales the difference may not be as large as suggested by the estimates, for example the estimate of those performing at Level 4/5 on the quantitative scale was 19% in England and 14% in Wales but the true proportion in England could be as low as 17% while the true proportion in Wales could be 15%. A recent survey in Wales using a different literacy assessment found 16% of the population aged 16-64 in Wales to have low or very low reading skills.[6]

(Figure 2.8)

As was shown earlier, literacy is related to a number of factors, such as age and education level, so this apparent difference between England and Wales could be related to differences in the distribution of

these characteristics in the population. In order to look at the differences between the countries it is necessary to hold other variables constant, for example to look at the proportion of people with higher education who are at each of the levels on the three literacy scales, and then to make a comparison between the countries. The rest of this section will look at literacy levels by age, educational attainment level and economic activity status for each of the three countries. Because of the limited sample size, some of the categories had to be combined.

2.12.1 Age

While the proportion of adults aged 16-25 who were at Level 4/5 on the prose scale was similar in England, Scotland and Wales the proportion of adults in the older age-groups who performed at this level was lower in Wales. Less than one in ten (9%) adults aged 26-45 living in Wales were at Level 4/5 compared with 21% and 14% of the same age-group living in England and Scotland respectively. This was the case both for men and women. There was a similar pattern on quantitative and document literacy with little difference in the proportion of the younger age-group at Level 4/5 across all three countries while the proportion of older people at Level 4/5 was lower in Wales than it was in England or Scotland.

(Figure 2.9, Table A2.8)

2.12.2 Educational attainment

On all three scales a smaller proportion of those with higher education living in Wales performed at the higher literacy levels than those with similar

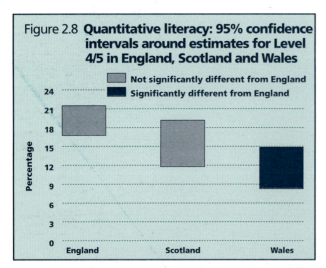

Figure 2.8 **Quantitative literacy: 95% confidence intervals around estimates for Level 4/5 in England, Scotland and Wales**

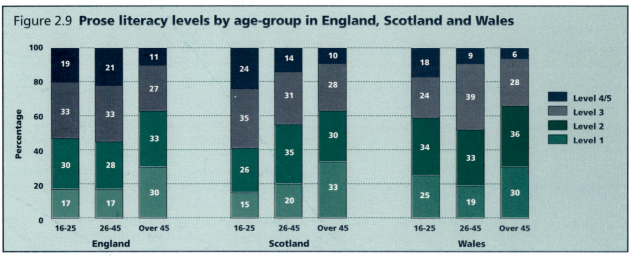

Figure 2.9 **Prose literacy levels by age-group in England, Scotland and Wales**

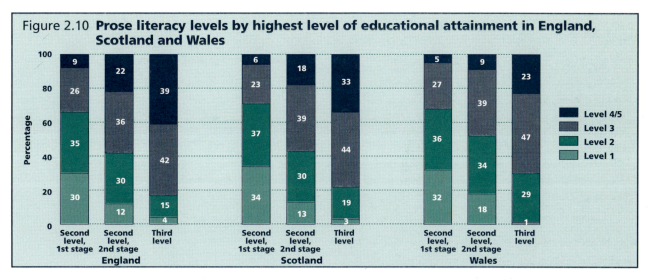

Figure 2.10 **Prose literacy levels by highest level of educational attainment in England, Scotland and Wales**

education but living in either England or Scotland. While 39% of those with higher education (ISCED 5 to 7) living in England and 33% of those with higher education living in Scotland were at Level 4/5 on prose only 23% of those with higher education living in Wales were at Levels 4/5 on this scale.

(Figure 2.10, Table A2.9)

2.12.3 Economic activity status

Adults in Wales who were in employment were less likely than those in employment in England or Scotland to be at Level 4/5 on any of the three dimensions of literacy. On prose literacy 9% of the employed were at Level 4/5 compared with 20% and 17% of the employed in England and Scotland respectively.

(Figure 2.11, Table A2.10)

Even when sex, age, education level and employment status are held constant Wales has a smaller proportion of people performing at the higher literacy levels (Level 4/5) on each of the three dimensions. Among the very youngest age-group in each country a similar proportion were at the highest literacy levels suggesting that the observed differences in the overall distribution of literacy in the population of working age may disappear over time.

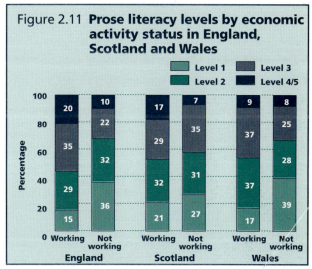

Figure 2.11 **Prose literacy levels by economic activity status in England, Scotland and Wales**

Notes and references

1. When controlling for educational attainment level, those in the oldest age-groups with the highest education levels while still more likely to be at the higher literacy levels than those in the same age-groups with lower educational levels, performed less well than younger people with similar levels of education suggesting that there is some effect due to age.

2. For a full description see *ISCED Handbook - United Kingdom (England and Wales), CSR/E/12,* UNESCO, Office of Statistics, December 1975. See also Bradley Mike et al. *England and Wales Youth Cohort Study Research studies RS10* DfEE

3. For the purpose of the British IALS the choice of which ISCED classification to use was determined by the need to provide weighted data for scaling of the literacy assessment, grossed up to population estimates and taking account of any non-response bias. International comparisons of educational attainment, whichever classification is used, are often based on information from the Labour Force Survey (LFS). As education is known to be highly correlated with literacy it was important to include education as one of the weighting variables. The LFS data file only contained one version

of ISCED, the version which shows the majority of the population at ISCED level 2 and this is therefore the version that was used. Once we received the literacy proficiency estimates it was possible to examine the relative merits of both versions of ISCED for future surveys of this type. The literacy proficiencies of those who would move from ISCED 2 to ISCED 3 between versions are more like those of people with ISCED 3 education than those with ISCED 2 education. The effect of using the ISCED classification which attributes these people to ISCED 2 is to increase the performance estimates of both ISCED 2 and ISCED 3. The weighting and grossing procedures used are described in Appendix A.

4. ONS *Standard Occupational Classification Vol. 1: Structure of the classification* HMSO (London 1990) and ONS *Standard Occupational Classification Vol. 3: Coding methodology* HMSO (London 1991).

5. The quintile bands were defined based on the personal income data of individuals aged 16-65 from the General Household Survey. The quintiles were grouped as follows based on annual yearly gross income from all sources: Quintile 1 - up to £2,704, Quintile 2 - £2,705-£5,928 Quintile 3 - £5,929-£10,400, Quintile 4 - £10,401-£16,848, Quintile 5: £16,849 or more.

6. The Basic Skills Agency. *Literacy & numeracy skills in Wales*. London: The Basic Skills Agency (1997).

3 Literacy and work

3.1 Introduction

It has already been seen that those currently in employment performed at higher skill levels than all other groups, except students, on all three literacy dimensions (Chapter 2). This chapter looks further at the relationship between literacy skills and work and examines literacy in the context of different occupations and industries as well as looking at occupational demands for literacy in the workplace. Analyses are based on the employed population plus those who had worked at any time in the 12 months prior to the survey as their recent employment may still contribute to their current literacy proficiency.

3.2 Literacy skill and industry

Much of the interest in the International Adult Literacy Survey comes from concern over declining employment in sectors that have traditionally provided low skill jobs in developed countries[1]. Many countries face problems of maintaining employment levels in traditional industries in the face of competition from emerging economies. Technological advances and increased automation have meant a decline in the number of low-skill jobs available as well as making certain skills/occupations obsolete. In addition entry requirements into employment continue to be affected by skill inflation where the skill threshold for employment at all levels continues to rise so that those with low educational qualifications find it increasingly difficult to find

employment. Also, jobs created in new or growing sectors tend to demand higher skill levels. Against this background, emphasis is increasingly being placed on the development of a highly skilled labour force; in particular the need for transferable skills and continuous skill development over an individual's working life. Literacy skills are a key tool in acquiring new skills and in taking advantage of opportunities for occupational change or development[2].

Figure 3.1 shows the percentage change in employment[3] since 1983 by industry and the average prose literacy proficiency scores for people currently employed or recently employed in those industries. In general, the industries that have seen the greatest increase in the number of employees over recent years, such as the financial sector and the sector that includes computers and research, have employees with high average literacy scores. Employees in industries that have seen the greatest decline in employment, such as agriculture/mining or construction, have lower average literacy scores than employees in most other sectors. This is of interest for two reasons. Firstly, if the numbers employed in those industries continue to decline, the skill levels of those released into the labour market will be below the average literacy proficiency required for the sectors which have experienced growth in employment. Secondly, even if employment in those industries does not contract any further, then the rate of growth of the high skill industries may mean that demand will outstrip the supply of high-skilled labour. In such

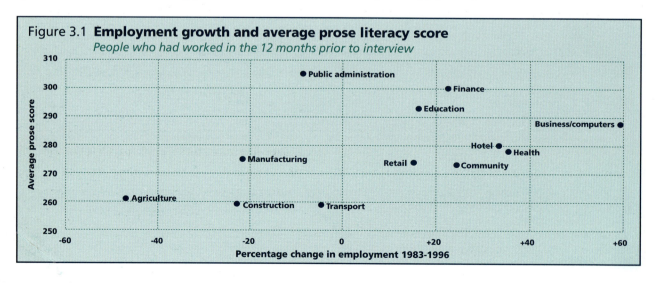

Figure 3.1 Employment growth and average prose literacy score
People who had worked in the 12 months prior to interview

circumstances it is unlikely that the demand for high skills can be met by new entrants to the labour market alone. The literacy skill levels of the reserve labour force (the unemployed and economically inactive) would also be insufficient to meet demand, because, as we have seen in Chapter 2, they are currently below the average proficiency levels of employees in those industrial sectors that have experienced growth.

(Figure 3.1, Table A3.1)

3.3 Literacy skill and occupation

In this chapter results are reported using the Standard Occupational Classification for Britain[4]. Classifications of occupation are usually based on assumptions about entry requirements into occupations such as skill levels, educational qualifications and experience. Given that the classification implies a hierarchy of skill it is not surprising then that the literacy skills of those in different occupational groups[5] varied quite considerably. As Table 3.1 shows, those in managerial, professional or technical occupations were more likely to perform at the higher literacy levels than those in other occupations. About a third of associate professional and technical occupations performed at Level 4/5 on prose literacy with higher proportions performing at that level for both the document and quantitative dimensions. Workers in clerical and secretarial occupations were more evenly distributed over the literacy skill levels with some workers at every level, although few (11 - 14%) were at Level 1. Workers in personal and protective service occupations, sales and skilled engineering occupations were more likely to perform at the middle literacy levels, Levels 2 or 3, while those in occupations such as machine operators and other elementary occupations were more likely to be at the two lowest literacy levels, Levels 1 and 2.

(Table 3.1, Table A3.2)

Respondents in some occupational groups were very consistent in their performance across the three literacy dimensions. Clerical and secretarial workers, for example, had almost identical distributions on prose, document and quantitative literacy.

Respondents in some other occupational groups performed better on one dimension than on others, for example, 39% of professionals were at Level 4/5 for prose, but 47% performed at that level for document literacy as did almost half (49%) for quantitative.

(Table 3.1, Table A3.2)

3.3.1 Occupation and educational attainment level

As mentioned above, educational attainment is one of the factors taken into account in the classification of occupations. The association between literacy level and occupation may simply be due to an underlying association with education. Table 3.2 shows the distribution of educational attainment for each of the occupational groups. The distribution matches very closely the pattern observed in Table 3.1. Some occupational groups are homogenous, such as professional occupations where large proportions are educated to university degree level or higher. Others are quite diverse, for example, managers who have substantial proportions at lower secondary, upper secondary and third level. Two occupational groups stand out as being most likely to have completed at least a university degree or its equivalent: science, engineering and health professionals and other professional groups (57% and 59% respectively). A further 21% of both these professional groups had third level education below degree level, for example, a Higher National Diploma (HND). The group with the next highest level of educational attainment were workers in the other associate professional occupations, with over a third (37%) having a third level education to degree level and a further 9% to below degree level. Among associate professionals in science, engineering and health 46% had tertiary education below degree level and 13% to degree level or higher. Over half of workers (53%) in the skilled engineering trades completed upper secondary education, for example to A-level or the vocational equivalent. The remaining occupation groups had a majority of respondents whose highest level of education was the completion of lower secondary school.

(Table 3.2)

Table 3.1 Literacy level by occupation
People who had worked in the 12 months prior to interview

	Level 1	Level 2	Level 3	Level 4/5	Total	Mean score	Base
	%	%	%	%	%		
Prose literacy							
Managers and administrators	8	28	39	25	100	291	*391*
Corporate managers and administrators	5	24	39	32	100	300	*260*
Managers/proprietors in agriculture and services	14	35	38	13	100	274	*131*
Professional occupations	3	15	43	39	100	312	*356*
Science engineering and health professionals	2	18	47	33	100	309	*104*
Other professional occupations	4	13	42	41	100	313	*252*
Associate professional and technical occupations	7	16	45	33	100	302	*282*
Science, engineering and health associate professionals	6	14	48	32	100	302	*146*
Other associate professional occupations	7	17	43	34	100	302	*136*
Clerical and secretarial occupations	11	29	38	23	100	286	*418*
Craft and related occupations	24	39	31	6	100	255	*299*
Skilled engineering trades	9	38	46	8	100	272	*92*
Other skilled trades	31	40	25	5	100	247	*207*
Personal and protective service occupations	18	29	36	18	100	276	*314*
Sales occupations	10	31	36	23	100	285	*220*
Plant and machine operators	32	41	20	6	100	246	*280*
Other occupations	35	36	23	6	100	244	*243*
Total	**16**	**29**	**35**	**20**	**100**	**278**	*2803*
Document literacy							
Managers and administrators	9	24	38	29	100	295	*391*
Corporate managers	5	18	42	35	100	307	*260*
Managers/proprietors in agriculture and services	16	35	32	17	100	275	*131*
Professional occupations	4	12	38	47	100	319	*356*
Science engineering and health professionals	3	12	40	46	100	322	*104*
Other professionals	4	11	37	47	100	318	*252*
Associate professional and technical occupations	8	17	40	35	100	303	*282*
Science, engineering and health associate professionals	6	16	45	34	100	302	*146*
Other associate professionals	10	19	35	36	100	304	*136*
Clerical and secretarial occupations	13	26	36	25	100	286	*418*
Craft and related occupations	21	37	30	12	100	263	*299*
Skilled engineering trades	9	30	45	16	100	285	*92*
Other skilled trades	27	39	23	10	100	253	*207*
Personal and protective service occupations	20	26	37	17	100	274	*314*
Sales occupations	16	25	35	24	100	282	*220*
Plant and machine operators	29	39	23	9	100	250	*280*
Other occupations	35	28	30	7	100	241	*243*
Total	**17**	**26**	**34**	**23**	**100**	**280**	*2803*
Quantitative literacy							
Managers and administrators	8	19	41	31	100	300	*391*
Corporate managers	3	16	40	41	100	313	*260*
Managers/proprietors in agriculture and services	16	27	43	14	100	277	*131*
Professional occupations	3	13	35	49	100	320	*356*
Science engineering and health professionals	2	10	37	51	100	326	*104*
Other professionals	4	14	34	48	100	318	*252*
Associate professional and technical occupations	7	19	38	37	100	306	*282*
Science, engineering and health associate professionals	6	20	43	31	100	302	*146*
Other associate professionals	8	17	33	42	100	310	*136*
Clerical and secretarial occupations	14	26	38	22	100	283	*418*
Craft and related occupations	21	38	26	14	100	265	*299*
Skilled engineering trades	12	32	37	19	100	286	*92*
Other skilled trades	25	41	22	12	100	256	*207*
Personal and protective service occupations	22	33	33	12	100	266	*314*
Sales occupations	19	27	34	21	100	277	*220*
Plant and machine operators	26	39	25	11	100	254	*280*
Other occupations	37	35	23	6	100	239	*243*
Total	**17**	**28**	**33**	**23**	**100**	**280**	*2803*

Table 3.2 Educational attainment level by occupation
People who had worked in the 12 months prior to interview

	Primary education or lower	Second level, 1st stage	Second level, 2nd stage	Third level non-university	Third level, university	Total	Base
	%	%	%	%	%	%	
Managers and administrators	2	42	23	11	22	100	*391*
Corporate managers and administrators	0	38	22	9	30	100	*260*
Managers/proprietors in agriculture and services	5	50	24	14	7	100	*131*
Professional occupations	0	9	11	21	58	100	*356*
Science, engineering and health professionals	-	7	16	21	57	100	*104*
Other professionals	1	10	9	21	59	100	*252*
Associate professional and technical occupations	1	30	17	26	26	100	*282*
Science, engineering and health associate professionals	1	26	14	46	13	100	*146*
Other associate professional occupations	1	34	18	9	37	100	*136*
Clerical and secretarial occupations	2	68	18	6	6	100	*418*
Craft and related occupations	3	49	40	5	3	100	*299*
Skilled engineering trades	2	37	53	7	1	100	*92*
Other skilled trades	3	54	35	5	3	100	*207*
Personal and protective service occupations	6	60	26	5	3	100	*314*
Sales occupations	8	57	26	5	4	100	*220*
Plant and machine operatives	2	74	21	1	2	100	*280*
Other occupations	11	72	14	2	2	100	*243*
Total	4	52	22	9	14	100	*2803*

Figure 3.2 looks more directly at the relationship between literacy levels, occupation group and educational attainment. By grouping professional and associate professional occupational groups together it is possible to look at those whose highest level of education is lower secondary level or lower and those educated to upper secondary level. When educational attainment is held constant much of the differences in literacy skill between occupations disappears. This is not, however, true for all occupational groups. Figure 3.2 shows the percentage of each occupational group at prose Levels 1 and 2 for each of the two educational groups. Higher proportions of craft workers, plant and machine operatives and workers in other occupations were at prose Levels 1 or 2 than were workers in other occupational groups with the same education. For example, among craft workers whose highest level of education was at lower secondary level or lower 34% were at prose Level 1 and a further 40% were at prose Level 2. In contrast, among those in clerical and secretarial occupations with the same level of education, 14% were at prose Level 1 and a third were at Level 2. The differences between these two occupational groups was even greater among those whose highest level of education was at upper secondary level.

(Figure 3.2, Table A3.3)

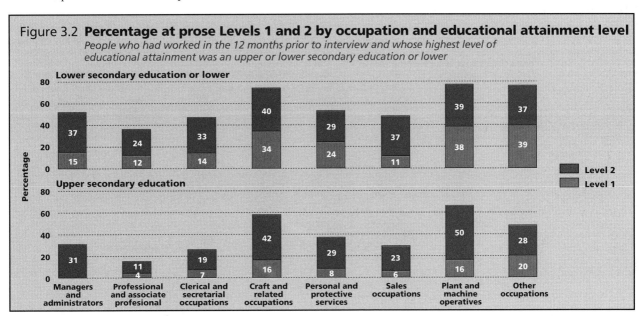

Figure 3.2 **Percentage at prose Levels 1 and 2 by occupation and educational attainment level**
People who had worked in the 12 months prior to interview and whose highest level of educational attainment was an upper or lower secondary education or lower

Table 3.3   Quantitative literacy level by whether works mostly full time or part time and sex
People who had worked in the 12 months prior to interview

	Level 1	Level 2	Level 3	Level 4/5	Total	Mean score	Base
	%	%	%	%	%		
Quantitative literacy							
Men							
Full time	14	26	31	30	100	289	1226
Part time	9	31	40	21	100	288	97
Total	13	26	31	29	100	283	1323
Women							
Full time	20	27	36	17	100	272	708
Part time	23	30	34	14	100	267	607
Total	21	28	35	15	100	271	1315
All							
Full time	16	26	32	26	100	283	1934
Part time	20	30	35	15	100	271	704
Total	17	27	33	23	100	280	2638

3.4 Literacy skills of full-time and part-time workers

Those who were employed on a part-time basis generally had the same literacy skills as those who worked full time except for the quantitative scale where there was a difference at the highest literacy level (Level 4/5). Over a quarter (26%) of those working full time were at Level 4/5 on quantitative literacy compared with just 15% of part-time workers. However, part-time workers are mainly women (82%) and Table 3.3, showing the distribution of quantitative literacy by full-time and part-time employment for men and women separately, demonstrates that most of the differences were due to gender. Among men, there was no significant difference in the literacy profiles of those in full-time versus part-time employment.

(Table 3.3, Table A3.4)

3.5 Occupational demands for literacy skills

Respondents were asked how often they had to perform different types of reading, writing and mathematical activities as part of their job although no attempt was made to determine the difficulty of these tasks. Conceivably, any of the activities could be at any level of difficulty. For example, reading a letter or memo could be very simple and represent a task at Level 1 or extremely complex and represent a Level 5 task.

Tables 3.4 and 3.5 show that all occupations involve activities which require the use of reading, writing and mathematical skills, although some occupations require them a lot less frequently than others. Managers, professionals and associate professionals were all regularly required to perform most of the reading tasks in contrast to the relatively low demand for workplace reading in plant and machine operative and other elementary occupations. Some tasks are more predominant in certain occupations, for example, reading diagrams is more frequently undertaken by professionals and skilled craft workers than by some other occupational groups. Reading directions or instructions for medicine, recipes or other products is an activity carried out more frequently by workers in personal service and associate professional occupations than other occupational groups. Reading or using information from a computer is more commonly required in managerial, professional and associate professional occupations with only small percentages of those in some other occupations being required to do so. What is clear, however, is that even in those occupations with poor average proficiency levels on the three literacy dimensions, substantial proportions of workers are regularly required to undertake activities that require these skills. For example, among plant and machine operators, a third of whom were at Level 1 on prose and a further 41% at Level 2, over half of the respondents in these occupations were required to read letters or memos at least once a week as part of their job and over a third were required to read reports, articles magazines or journals, or manuals/reference books or read or use information from diagrams.

(Table 3.4)

Table 3.5 shows a similar pattern for writing and mathematical tasks with certain occupations performing them more regularly than others. In general, a lower percentage of each occupational group reported involvement in writing and mathematical tasks than they did for reading. Two notable exceptions to this are workers in craft and related occupations and machine operatives. Over two thirds (69%) of craft workers and almost half (49%) of plant and machine operatives measure or estimate the size or weight of objects at least once a week as part of their job.

(Table 3.5)

3.6 Literacy skill and literacy practice at work

The relationship between literacy skill and literacy practice at work is unclear and is complicated by the fact that some occupations require more frequent use of literacy skills than others. Generally, those who report engaging in literacy activities at least once a week as part of their job demonstrated higher literacy skills on each dimension than those who reported engaging in these activities less frequently. For example, 13% of those who read memos or letters at least once a week were at Level 1 on the document scale compared with over a quarter (26%) of those who are not involved in this activity on at least a weekly basis.

(Table 3.6)

It has already been shown that those in occupations with high average literacy are more likely to be

Table 3.4 Percentage of workers in each occupational group who reported engaging in each of several workplace reading tasks at least once a week
People who had worked in the 12 months prior to interview

	Letters or memos	Reports, articles, magazines or journals	Manuals or reference books, including catalogues	Diagrams	Bills, invoices, spreadsheets or budget tables	Directions or instructions for medicines, recipes or other products	Use information from computers	Base
	%	%	%	%	%	%	%	
Managers and administrators	93	85	72	48	85	38	73	388
Professional occupations	94	90	82	60	49	37	79	353
Associate professional and technical occupations	93	88	78	46	53	45	76	280
Clerical and secretarial occupations	91	65	60	23	65	18	83	413
Craft and related occupations	63	40	52	59	35	19	32	292
Personal and protective service occupations	59	47	37	22	26	51	27	305
Sales occupations	63	49	51	27	48	23	47	211
Plant and machine operatives	54	36	35	36	25	23	34	263
Other occupations	33	19	13	15	15	14	10	234
Total	73	59	54	37	47	30	54	2739

Table 3.5 Percentage of workers in each occupational group who reported engaging in each of several workplace writing and mathematical tasks at least once a week
People who had worked in the 12 months prior to interview

	Writing tasks				Mathematical tasks		Base
	Letters or memos	Forms or things such as bills, invoices or budgets	Reports or articles	Estimates or technical specifications	Measure or estimate the size or weight of objects	Calculate prices, costs or budgets	
Managers and administrators	82	78	57	35	47	82	388
Professional occupations	87	58	62	29	46	45	353
Associate professional and technical occupations	76	53	59	32	45	44	280
Clerical and secretarial occupations	73	61	32	17	30	51	413
Craft and related occupations	41	49	28	32	69	38	292
Personal and protective service occupations	42	35	36	7	35	32	305
Sales occupations	38	50	17	14	47	68	211
Plant and machine operatives	25	35	21	7	49	22	263
Other occupations	11	21	9	4	23	12	234
Total	55	51	37	20	43	45	2739

involved in such tasks as reading letters or memos. When looking at an activity which is less predominantly associated with particular occupations, such as reading directions or instructions for medicine, recipes or other products (Table 3.4), a relationship between frequent skill use and demonstrated skill is still evident. A significantly higher proportion of those who are not required to use these items at least weekly as part of their job performed at Level 1 on the document scale than did those who used them regularly. For example, among those who read reports or articles less than once a week 25% were at Level 1 (document) compared to only 11% of those who read them at least once a week. However, this is not necessarily to imply that engaging in these activities at work leads to high literacy. Large proportions of those with low

demonstrated literacy proficiency are required to undertake literacy tasks as part of their job, albeit less frequently than those with higher literacy skills. Over half (58%) of those at Level 1 on document literacy read letters and memos in their job at least once a week and more than a third are required to read information from reports or articles (39%), from manuals and reference books (36%) and from bills invoices, spreadsheets and budgets (35%). So performing the task regularly does not in itself lead to high skill.

(Table 3.6, Table 3.7 and Tables A3.5, A3.6, A3.7, A3.8)

Table 3.6 Document literacy level by frequency of engaging in several reading activities at work

People who had worked in the 12 months prior to interview

		Level 1	Level 2	Level 3	Level 4/5	Total	Mean score	Base
		%	%	%	%	%		
Document literacy								
Letters or memos	At least once a week	13	24	36	27	100	289	2056
	Less than once a week	26	31	30	13	100	257	701
Reports, articles, magazines or journals	At least once a week	11	24	36	29	100	294	1690
	Less than once a week	25	29	31	14	100	261	1067
Manuals or references books, including catalogues	At least once a week	11	22	37	29	100	294	1527
	Less than once a week	23	31	31	15	100	263	1229
Diagrams	At least once a week	11	23	38	28	100	294	1058
	Less than once a week	20	28	32	20	100	272	1699
Bills, invoices, spreadsheets or budget tables	At least once a week	12	24	36	27	100	290	1266
	Less than once a week	21	28	33	19	100	272	1491
Directions or instructions for medicines, recipes or other products	At least once a week	16	23	39	22	100	285	830
	Less than once a week	17	27	33	23	100	278	1926
Use information from computers	At least once a week	9	21	39	32	100	299	1497
	Less than once a week	26	32	29	13	100	258	1260
Total		17	26	34	23	100	280	2757

Table 3.7 Percentage of respondents at each document literacy level who reported engaging in each of several reading activities at work at least once a week

People who had worked in the 12 months prior to interview

	Letters or memos	Reports, articles, magazines or journals	Manuals or reference books, including catalogues	Diagrams	Bills, invoices, spreadsheets or budget tables	Directions or instructions for medicines, recipes or other products	Use information from computers	Base
	%	%	%	%	%	%	%	
Document literacy								
Level 1	58	39	36	24	35	28	29	470
Level 2	68	54	46	33	43	26	42	736
Level 3	77	62	59	41	50	33	60	932
Level 4/5	85	74	70	46	56	29	74	619
Total	73	59	54	37	47	30	54	2757

3.7 Self-assessment of skills for work

Respondents were asked to evaluate their reading, writing and mathematics skills for their job and whether they felt their skills were limiting their future job opportunities in any way. Overall very few people rated either their reading, writing or mathematical skills as poor for their job. Not surprisingly, most of those at literacy Level 3 or higher rated their skills as either good or excellent. One area of interest is the proportion of those at the lower literacy levels who considered their skills as inadequate for their job. One in ten of those at Level 1 on prose rated their reading skills as poor for their job and a further 30% rated their skills as moderate. Among those at Level 2 only one in five respondents considered their reading to be either poor (2%) or moderate (18%). This does not take into account whether these skills are required in their work. Those at Level 1 for example who rate their reading skills as either good (41%) or excellent (13%) may do so because reading is not something they are required to do regularly as part of their job therefore their current skill level is sufficient for their job requirements.

People generally rated their reading skills higher than either their writing or mathematics skills. Almost half of those at Level 1 on the prose scale considered their writing skills to be either moderate (35%) or poor (13%). Similarly, 12% of those at Level 1 on quantitative literacy rated their mathematics skills as poor. On the quantitative scale a small percentage of those who were at Level 3 or Level 4/5 rated their skills as poor. Although they demonstrated high levels of competence on the literacy assessment they may be in occupations which demand high numeracy skills or specialist mathematical ability.

(Table 3.8 and Tables A3.9, A3.10, A3.11)

Only a small proportion of respondents felt that their reading, writing or mathematics skills were limiting their job opportunities - either job advancement or getting another job. Those at the lower literacy levels were more likely to consider their skills as limiting their opportunities. Among those at Level 1 on the prose scale, 8% thought their reading skills greatly limiting and 20% considered them to be somewhat limiting their job opportunities. Respondents were less likely to consider their poor reading skills as limiting their job opportunities than either their

Table 3.8	Self-assessment of reading, writing and mathematics skills for main job by literacy level

People who had worked in the 12 months prior to interview

	Excellent	Good	Moderate	Poor	No opinion	Total	Base
	%	%	%	%	%	%	
Self-assessment of reading skills for main job							
Prose literacy							
Level 1	13	41	30	10	6	100	*434*
Level 2	26	53	18	2	2	100	*816*
Level 3	46	45	6	0	2	100	*960*
Level 4/5	68	28	3	-	1	100	*548*
Total	39	44	13	2	2	100	*2758*
Self-assessment of writing skills for main job							
Prose literacy							
Level 1	10	36	35	13	5	100	*434*
Level 2	17	51	27	3	2	100	*814*
Level 3	33	51	12	1	2	100	*960*
Level 4/5	51	41	6	0	2	100	*548*
Total	28	47	19	4	2	100	*2756*
Self-assessment of mathematics skills for main job							
Quantitative literacy							
Level 1	8	32	42	12	6	100	*477*
Level 2	15	49	28	4	3	100	*744*
Level 3	28	52	14	1	4	100	*914*
Level 4/5	45	46	7	1	2	100	*622*
Total	25	46	21	4	4	100	*2757*

Table 3.9   Self-assessment of whether reading, writing and mathematics skills are limiting job opportunities by literacy level

People who had worked in the 12 months prior to interview

	Greatly limiting	Somewhat limiting	Not at all limiting	Total	Base
	%	%	%	%	
Reading skills limiting job opportunities					
Prose literacy					
Level 1	8	20	72	100	431
Level 2	2	13	85	100	810
Level 3	1	6	94	100	959
Level 4/5	0	2	97	100	548
Total	2	9	88	100	2748
Writing skills limiting job opportunities					
Prose literacy					
Level 1	8	29	63	100	430
Level 2	2	19	80	100	813
Level 3	1	8	91	100	957
Level 4/5	-	2	98	100	548
Total	2	13	85	100	2748
Mathematical skills limiting job opportunities					
Quantitative literacy					
Level 1	10	30	60	100	476
Level 2	2	16	82	100	740
Level 3	0	9	91	100	913
Level 4/5	0	5	95	100	621
Total	2	14	84	100	2750

writing or mathematical skills. Again using the prose scale as the example, 37% of those at Level 1 considered their writing skills to be either greatly limiting (8%) or somewhat limiting (29%). Among those at Level 1 on the quantitative scale 10% considered their mathematics skills as greatly limiting and 30% as somewhat limiting their job opportunities. Large proportions of those at the lower literacy levels on all dimensions did not consider their skills to limit their job opportunities in any way.

(Table 3.9 and Tables A3.12, A3.13, A3.14))

3.8 Literacy skill and adult education and training

It is anticipated that in the future most workers will change occupation several times in their working life in response to technological development and skills obsoletion. Such occupational flexibility will require continuous development of existing skills and acquisition of new skills while in employment. An important part of the development of such flexibility is access to training and education opportunities. All respondents were asked whether, in the 12 months

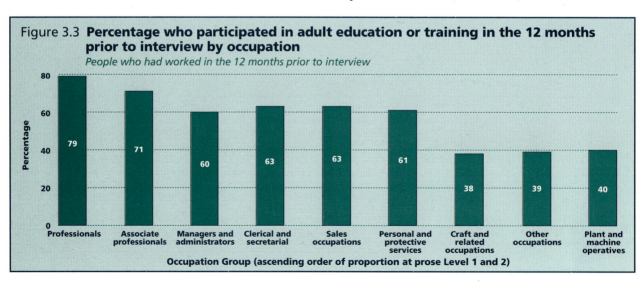

Figure 3.3 **Percentage who participated in adult education or training in the 12 months prior to interview by occupation**

People who had worked in the 12 months prior to interview

prior to interview, they had received any training or education[6] either for work or leisure. Over half (57%) of those who had worked in the 12 months prior to interview had participated in some form of education or training during the previous 12 month period. Figure 3.3 shows that those in occupations with low average literacy skill were least likely to have received any education or training in the reference period. Forty per cent or less of those in craft and related occupations, plant and machine operatives and other occupations had taken part in any education or training in the previous 12 months compared with 79% of professionals, 71% of associate professionals and between 60 and 63% of those in the remaining occupational groups.

(Figure 3.3, Table A3.15)

Table 3.10 shows, for some occupation groups[7] only, the proportion at each literacy level (Levels 3 and higher combined) who participated in adult education or training in the 12 months prior to interview. For most occupations those at the higher literacy levels (Level 3 or higher) are more likely to have received education or training than those at the two lowest literacy levels. Among clerical workers for example, of those at Level 3 or higher on document, 73% had participated in adult education or training in the 12 months prior to interview compared to only 43% of those at Level 1. Over half (54%) of those in craft and related occupations who were at literacy Level 3 or higher had taken part in some training or education activity compared with 21% of those in the same occupation who were at literacy Level 1. The differences between literacy levels in the rates of participation in adult education or training were less marked in some occupations, such as plant and machine operatives and sales occupations. The reason(s) why those with lower literacy skills do not take part in training and education opportunities as frequently as those with higher skills may be because people with low skills avoid situations, such as training courses, where their poor skills will be exposed, or it may be that employers do not offer training opportunities to those with low skills. Whatever the reason, the low participation rate of those with low literacy skills in training and education opportunities,

whether through self exclusion or through lack of opportunity may further polarise the distribution of literacy skills.

(Table 3.10)

Table 3.10	**Percentage of workers at each literacy level in selected occupational groups* who participated in adult education or training in the last 12 months**		
People who had worked in the 12 months prior to interview			
	Level 1	Level 2	Level 3/4/5
	Percentage who participated in adult education or training		
Prose literacy			
Clerical and secretarial occupations	33	53	74
Craft and related occupations	18	31	60
Personal and protective service occupations	34	52	74
Sales occupations	[12]	51	73
Plant and machine operatives	38	35	49
Other occupations	18	36	67
Document literacy			
Clerical and secretarial occupations	43	51	73
Craft and related occupations	21	31	54
Personal and protective service occupations	34	46	77
Sales occupations	46	52	72
Plant and machine operatives	38	34	50
Other occupations	21	31	62
Quantitative literacy			
Clerical and secretarial occupations	47	53	72
Craft and related occupations	22	34	51
Personal and protective service occupations	36	55	77
Sales occupations	61	51	69
Plant and machine operatives	39	39	42
Other occupations	20	41	59
Bases			
Prose literacy			
Clerical and secretarial occupations	*48*	*125*	*245*
Craft and related occupations	*71*	*118*	*110*
Personal and protective service occupations	*65*	*98*	*151*
Sales occupations	*26*	*73*	*121*
Plant and machine operatives	*91*	*110*	*79*
Other occupations	*94*	*87*	*62*
Document literacy			
Clerical and secretarial occupations	*54*	*112*	*252*
Craft and related occupations	*67*	*104*	*128*
Personal and protective service occupations	*73*	*98*	*143*
Sales occupations	*39*	*62*	*119*
Plant and machine operatives	*86*	*105*	*89*
Other occupations	*99*	*73*	*71*
Quantitative literacy			
Clerical and secretarial occupations	*62*	*118*	*238*
Craft and related occupations	*65*	*107*	*127*
Personal and protective service occupations	*79*	*106*	*129*
Sales occupations	*45*	*64*	*111*
Plant and machine operatives	*78*	*99*	*103*
Other occupations	*103*	*81*	*59*

* Because of the small number of people in the high skill occupation groups that are at the lower literacy levels it is not possible to examine the relationship between literacy level and participation in training and education for these groups

3.9 Literacy skill and wage income

If the trend in developed countries is for the economy to be increasingly dependent on skills, then one would expect high skills to be rewarded. Chapter 2 looked at the relationship between personal income and literacy levels. However, this is not necessarily a good indication of an individual's earning power as some people have substantial income from other sources. This section looks at the wage income of those in full-time employment. Again, to aid comparison with other countries, annual gross income quintiles[8] are used. Figure 3.4 shows a clear relationship between literacy skills and income, with those with the greatest income from employment having the highest literacy skills. In all but the highest income quintile, similar proportions in each wage income quintile were at the middle literacy levels. The literacy profiles of those in the two lowest income quintiles were very similar. Among those in the lowest wage income quintile 38% were at Level 1 on quantitative literacy and a further 35% were at literacy Level 2. Those in the middle income range were most likely to be at the middle skill levels, a third (32%) were at Level 2 on quantitative and a further third (33%) were at Level 3. Almost half (47%) of those in the highest income quintile performed at Level 4/5 on quantitative literacy and a further 36% were at Level 3 with very few demonstrating low quantitative literacy ability.

(Figure 3.4, Table A3.16)

Income from employment is also related to age as the remuneration of new labour market entrants reflects their lack of experience. Among those aged 16-25 wage income and literacy skill were not as strongly associated as for other age-groups. Among those aged 26-45 and aged 46-65 the overall pattern held, with those at the higher literacy levels more likely to be at the higher wage income levels than those at the lower literacy levels.

(Figure 3.5, Table A3.17)

Notes and references

1   For example *Competitiveness, Occasional Paper The Skills Audit: A Report from an Interdepartmental Group.* Department for Education and Employment and Cabinet Office.   (1996 London).

2   See for example, Anderson, Alan and Marshall, Vivien. Core versus occupation-specific skills *Research Studies*

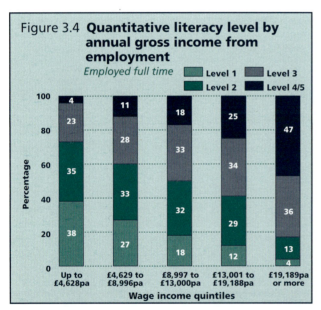

Figure 3.4 **Quantitative literacy level by annual gross income from employment**
Employed full time
Level 1   Level 3
Level 2   Level 4/5

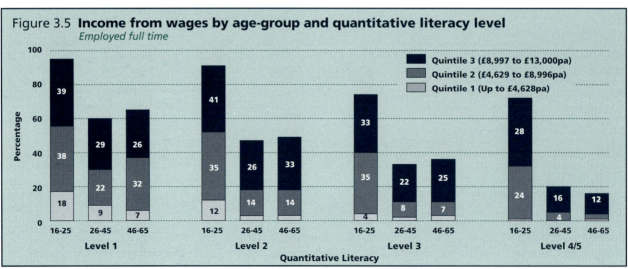

Figure 3.5 **Income from wages by age-group and quantitative literacy level**
Employed full time

Quintile 3 (£8,997 to £13,000pa)
Quintile 2 (£4,629 to £8,996pa)
Quintile 1 (Up to £4,628pa)

RS12. Department for Education and Employment. HMSO (1996 London).

3   The percentage changes in the number of employees in employment in Great Britain (seasonally adjusted), June 1983 to March 1996, have been calculated from the monthly bulletin of the Office for National Statistics *Labour Market Trends* April 1997, Table 1.2, page S9.

4   In the international survey, occupation was coded to the major groups of the international classification ISCO-88. Within some occupation major groups there are important differences in the literacy skills of various sub-groups, which otherwise might be obscured when the broad groupings are used. Therefore this chapter uses the British Standard Occupation Classification. In Chapter 6, which deals with international comparisons, the results for Great Britain are reported using ISCO-88.

5   Examples of some of the different jobs in each of the occupation groups used in this section can be found in the glossary of definitions and terms (Appendix B) under the heading Standard Occupation Classification.

6   This included training of any duration and included leisure courses as well as courses related to their work.

7   Because of the small number of people in the high skill occupation groups that are at the lower literacy levels it is not possible to examine the relationship between literacy level and participation in training and education for these groups.

8   The annual wage income quintiles for the working population of Great Britain, aged 16-65, were calculated from the income data collected by the 1994 General Household Survey (GHS). The annual wage income quintiles were 1. up to £4,628 pa, 2. £4,629-£8,996 pa, 3. £8,997-£13,000 pa, 4. £13,001-£19,188 pa, and 5. £19,189 pa or more. The questions did not take into account the number of hours worked.

4 Literacy in everyday life

4.1 Introduction

Chapter 3 looked at the use of literacy skills in the workplace but literacy skills are also used in everyday life. The ability to apply literacy skills in a wide range of different contexts is important. For example, people need to be able to read and understand medicine labels, bus and train timetables, recipes or instructions for consumer goods such as video recorders. This chapter presents information on literacy practices, such as reading books and newspapers or magazines, and activities related to literacy such as time spent watching television and looks at the relationship between these practices and literacy levels. It also looks at respondents' self-assessment of their literacy skills in everyday life.

4.2 Literacy practices

Newspapers or magazines were read by almost everyone, 72% of respondents read them daily and 22% once a week. Only 2% never read them. Newspapers and magazines vary in style and content, and hence in reading difficulty; out of the wide range of material available, there is therefore something that is accessible to people at all literacy levels. The majority (71%) of those who never read a newspaper or magazine were at prose Level 1 and only 4% at Level 4/5.

(Figures 4.1 and 4.2, Table A4.1)

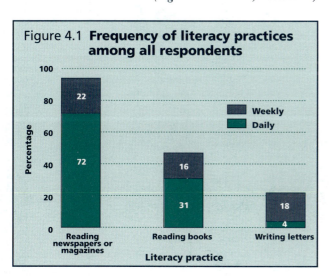

Figure 4.1 **Frequency of literacy practices among all respondents**

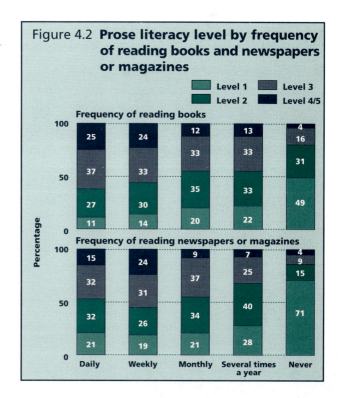

Figure 4.2 **Prose literacy level by frequency of reading books and newspapers or magazines**

A better indicator of respondents' literacy skills was the frequency of reading books. Thirty one percent of respondents reported reading a book every day and 16% once a week; 18% never read books. Those who read books daily were more likely than those who never read books to be at Levels 3 or above on all three literacy dimensions. For example, of those who read a book daily 37% were at Level 3 (prose) and 25% were at Level 4/5, compared to 16% and 4% respectively of those who never read a book.

(Figures 4.1 and 4.2, Table A4.2)

Writing is not an activity that is engaged in regularly by most people. Less than a quarter (22%) of respondents said they wrote letters or anything else more than a page in length at least once a week (excluding work) and 25% never wrote letters. Those who wrote letters weekly tended to be at the higher literacy levels (Level 4/5), 26% were at prose Level 4/5 and only 12% at Level 1. In contrast only 7% of those who never wrote letters were at Level 4/5 and 39% were at Level 1.

(Figure 4.1, Table A4.3)

4.3 Access to reading materials

Respondents were asked whether certain reading materials were present in their home. Ninety three percent of respondents said that there was a dictionary in their home, 83% had more than 25 books and 70% had a daily newspaper. The possession of a daily newspaper in a household did not appear to be related to literacy skills, although as suggested earlier this is probably due to the wide variety of newspapers and magazines available. However, having fewer than 25 books or not having a dictionary were associated with low literacy levels. Fifty one percent of respondents whose household did not have at least 25 books were at prose Level 1 whilst only 3% were at Level 4/5. Similarly among respondents whose household did not have a dictionary, 53% were at Level 1 prose and only 6% at Level 4/5. While the absence of these materials in the household may not be directly related to literacy level it does give an indication of the role of literacy in peoples' daily lives and also the extent to which people have the opportunity to develop their literacy skills at home.

(Figure 4.3, Tables 4.1, A4.4-A4.6)

Reading materials are also available outside the home, for example in public libraries. Although only 29% of respondents reporting visiting a public library at least once a month, those that did so were more likely to perform at the higher literacy levels; 37% of those who used a library at least once a month were at Level 3 (prose) and 24% were at Level 4/5 whereas only 12% were at Level 1. The majority of respondents who never used a library were either at Level 1 or Level 2 on all three dimensions.

(Figure 4.3, Tables 4.1, A4.7)

4.4 Literacy skills, television and current events

The survey asked respondents how much time they spent each day watching television. For many people this is an important means of getting information and keeping abreast of current events. Figure 4.4 shows

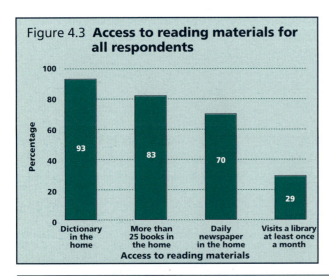

Figure 4.3 **Access to reading materials for all respondents**

Table 4.1	**Prose literacy level by access to reading materials**						
	Level 1	Level 2	Level 3	Level 4/5	Total	Mean score	*Base*
	%	%	%	%	%		
Whether respondent has more than 25 books in their home							
More than 25 books	16	30	34	19	100	277	*3130*
25 or fewer	51	30	16	3	100	215	*676*
Total	22	30	31	17	100	267	*3806*
Whether household has a dictionary							
Yes	19	30	33	17	100	271	*3477*
No	53	28	13	6	100	211	*325*
Total	22	30	31	17	100	267	*3802*
Whether household takes a daily newspaper							
Yes	21	31	32	16	100	267	*2629*
No	22	29	30	19	100	266	*1177*
Total	22	30	31	17	100	267	*3806*
Frequency of using a public library							
At least once a month	12	27	37	24	100	287	1164
Several times a year	13	29	38	20	100	283	987
Never	33	34	23	10	100	243	*1655*
Total	22	30	31	17	100	267	*3806*

that those watching the most television were more likely to be at lower literacy levels. Of those who watched more than 5 hours of television per day 42% were at Level 1 (prose) compared with only 5% at Level 4/5. In contrast almost a third (30%) of those who watched one hour or less per day were at Level 4/5 and 12% at Level 1. While there is an association between literacy level and the amount of time spent watching television, it does not necessarily imply that watching television causes low literacy skill, it is just as plausible that watching a lot of television is a consequence of having low literacy skills. Watching television may however reduce the time available for literacy based activities such as reading books if it is done at the exclusion of such activities. Table A4.9 shows that respondents who reported watching television for five hours or more per day were the least likely to read a book daily (23%) or weekly (10%) whereas over half of those who reported watching television for an hour or less per day read a book at least weekly (36% daily and a further 24% weekly).

(Figure 4.4, Tables A4.8 and A4.9)

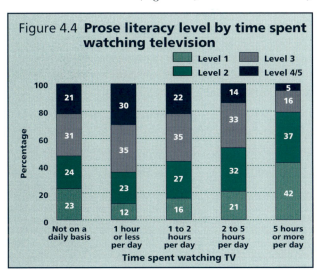

Figure 4.4 **Prose literacy level by time spent watching television**

Literacy is important, not only for the daily routine of our lives such as dealing with bills, letters from school and filling in forms, but also for keeping abreast of social and political events. Respondents were asked how often they followed what is going on in current events, government and public affairs. Of those who said they followed current events hardly at all 54% were at Level 1 compared with only 3% at Level 4/5 and 17% at Level 3. Although those who rarely

followed current events were predominantly low skilled, this is not to imply that all those with low literacy skills are marginalised from social or public affairs. Among those who regularly follow current events 16% were at document Level 1 and a quarter were at Level 2.

(Figure 4.5, Table A4.10)

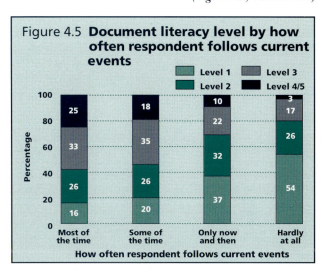

Figure 4.5 **Document literacy level by how often respondent follows current events**

The use of computers is now a common feature of many jobs and an increasing number of homes now have personal computers[1]. Respondents were asked how often they used a computer at home. In Chapter 3 we saw that workers who used computers regularly at work tended to have higher literacy skills than workers who did not and the same is true of computer use in the home. Of those who reported using a computer at home daily 30% were at Level 4/5 on the quantitative scale while only 8% were at Level 1.

(Table A4.11)

4.5 Self-assessment of literacy skill

Respondents were asked to rate their literacy skills for everyday life to see if those with low skills recognised their limitations in dealing with printed materials. Generally, respondents were very pleased with their literacy skills, with 46% considering their reading skills as 'excellent' and a further 40% describing them as 'good'. There was a slightly lower level of satisfaction with writing and mathematics, 33% rating their writing skills as 'excellent' and 45% rating them as 'good' and 25% rating their mathematics skills as 'excellent' and 45% rating them as 'good'.

(Figure 4.6, Table A4.12)

Comparing self-assessment with actual skill levels, as measured by the survey, those who considered their skill as either poor or moderate were more likely to be at Levels 1 or 2 on the literacy scales than those assessing their skills as either good or excellent. Almost all (86%) of those who said their reading skills were poor were at Level 1 with a further 8% at Level 2 and 6% at Level 3; none of those who described their reading skills as poor were at Level 4/5. However, almost 1 in 10 of those who said their reading skills were excellent and almost a quarter of those who said they were good were at prose Level 1.

(Figure 4.7, Tables A4.13-A4.15)

Although a small proportion of respondents assessed their reading, writing or mathematics skills for daily life as poor, the majority of those that did were at Level 1 on all three dimensions. The high overall level of satisfaction with literacy skill levels for everyday life may reflect the way people adjust their lives to match their skill levels or it may reflect the lack of demand placed on them to do literacy tasks. For example, people with low skills tend to avoid situations where they have to use skills and tasks they find difficult or perhaps do not find themselves in situations where they are expected to use high levels of literacy skill. If they have strategies for avoiding the need to use such skills and are never expected to use them, then the skills they do have may well appear adequate for their everyday life. The self-assessment of literacy skills by the low skilled and whether they need help with tasks such as reading or filling out forms is described in Chapter 5.

4.6 Literacy in the home and school-age children

The literacy skills of young people continues to be an issue of topical interest. Programmes to improve childrens' basic skills often try to encourage reading practice at home with the involvement of parents.[2] Research suggests that where parents have poor skills their children are also likely to have poor skills[3]. Respondents who had children aged 6 to 15 living with them, were asked about literacy in the home environment. Respondents at Level 1 were less likely than those at higher literacy levels to report that their children would often see the respondent or their partner reading. Almost all respondents (99%) at prose Level 4/5 said their children would often see them or their partner reading, as did almost all of those at Level 3 (96%) and Level 2 (89%) compared with 68% of those at Level 1. There was no consistent pattern between parents' literacy levels and whether or not parents said their children had time set aside for reading or were limited in the amount of time they were allowed to watch television.

(Table 4.2, Table A4.16)

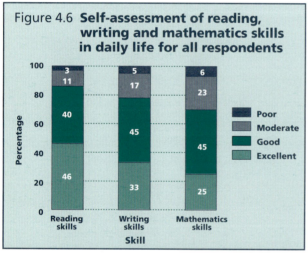

Figure 4.6 **Self-assessment of reading, writing and mathematics skills in daily life for all respondents**

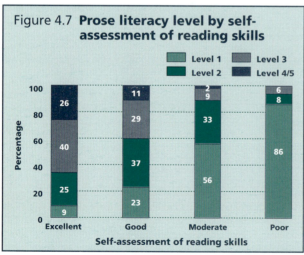

Figure 4.7 **Prose literacy level by self-assessment of reading skills**

Table 4.2 **Percentage of respondents who reported various family literacy practices**

People who had children aged 6 to 15 living with them

| | Children of respondent... | | | |
	often see respondent or spouse reading	have time set aside for reading	are limited in amount of time allowed to watch TV	*Base*
	Percentage of respondents			
Prose literacy				
Level 1	68	46	48	*173*
Level 2	89	55	55	*291*
Level 3	96	47	54	*309*
Level 4/5	99	52	46	*158*
Total	89	50	51	*931*

Notes and references

1. 25% of households in Great Britain in 1995 had a home computer compared to 13% in 1985 (General Household Survey 1995).

2. see for example Brooks, Greg et al. *Family Literacy Works.* London: Basic Skills Agency 1996.

3. Bynner J and Steedman, J. *Difficulties with Basic Skills: findings from the 1970 British Cohort Study.* London: Basic Skills Agency 1995.

5 People with low literacy skills

5.1 **Introduction**

While the main focus of this report deals with the literacy skills of the whole population aged 16-65 one of the groups of most interest are those who perform at the lower end of the ability range and, in particular, those performing at Level 1 which is the main focus of this chapter. Just over a fifth of adults aged 16-65 were at Level 1 on the prose scale and a similar proportion were at Level 1 on the document and quantitative scales. It was shown in Chapter 3 that industries with low average literacy proficiency such as agriculture, mining and construction are in decline while there is an increase in the number of jobs requiring higher literacy skills such as finance and computing. If people with low skills do not improve their levels of literacy they may find it increasingly difficult to obtain employment. It is, therefore, important to identify this low-skilled group and the chapter begins by providing a portrait of people at Level 1 and then going on to describe the characteristics of people at the other levels in order to consider the ways in which the characteristics of the low-skilled group differ from those of people with higher literacy levels. This chapter also looks at the low-skilled in terms of their literacy practices compared with high-skilled people and at their self-assessment of literacy skills. As there is a high correlation between performance on the prose,

document and quantitative literacy scales figures quoted in the text are for one scale only.

5.2 **The characteristics of people at Level 1**

People who perform at Level 1 can make limited use of texts that are simple and uncomplicated. They are able to locate information in text or data as long as there is no distracting information around the correct answer. On the quantitative scale they can carry out relatively straightforward operations such as simple addition. Those who performed at Level 1 on the three literacy scales were predominantly older people with low levels of education. They were more likely than people at higher levels to be unemployed, to belong to the manual rather than the non-manual social classes and to be on a low income. This is similar to the findings reported recently by The Basic Skills Agency[1]. Those at Level 1 on the document and quantitative scales (but not the prose scale) were more likely to be women.

- 19% of those at Level 1 on the prose scale were aged 46-55 and a further 28% were aged 56-65 meaning that 47% were aged over 45 compared with only 22% of those at Level 4/5.

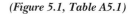

(Figure 5.1, Table A5.1)

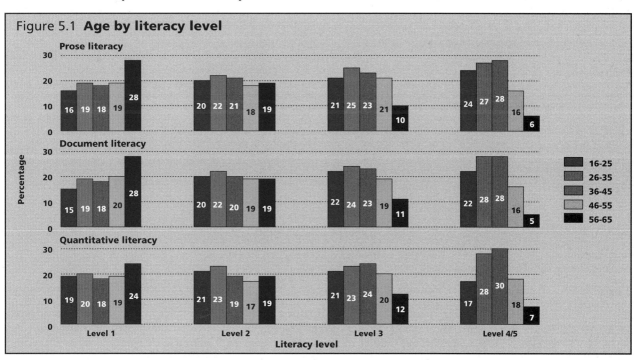

Figure 5.1 **Age by literacy level**

Table 5.1 Sex by literacy level

	Male		Female		Total	Base
	%	s.e	%	s.e	%	
Prose literacy						
Level 1	48	2.1	52	2.1	100	825
Level 2	49	2.0	51	2.0	100	1165
Level 3	52	2.2	48	2.2	100	1204
Levels 4/5	51	2.8	49	2.8	100	617
Total	**50**	**1.4**	**50**	**1.4**	**100**	3811
Document literacy						
Level 1	43	1.9	57	1.9	100	900
Level 2	47	1.8	53	1.8	100	1057
Level 3	51	2.5	49	2.5	100	1154
Levels 4/5	62	2.6	38	2.6	100	700
Total	**50**	**1.4**	**50**	**1.4**	**100**	3811
Quantitative literacy						
Level 1	38	2.1	62	2.1	100	905
Level 2	48	1.7	52	1.7	100	1070
Level 3	50	2.6	50	2.6	100	1146
Levels 4/5	69	2.9	31	2.9	100	690
Total	**50**	**1.4**	**50**	**1.4**	**100**	3811

s.e = Standard Error of the estimate. The reported sample estimate can be said to be within 2 standard errors of the true population value with 95% confidence.

Figure 5.2 Highest level of educational attainment by prose literacy level

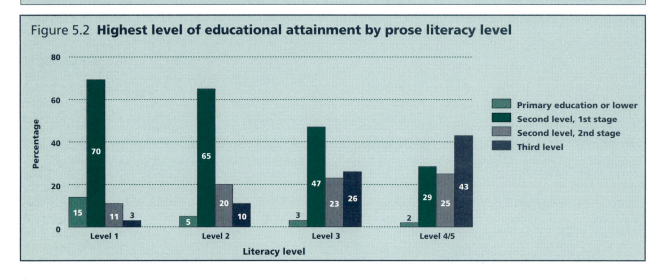

- 57% of those at Level 1 on the document scale were women as were 62% of those on the quantitative scale. There was no such preponderance of women among those performing at Level 1 on the prose scale.

  (Table 5.1)

- Most of those with low skills on the three scales had completed their education at lower secondary level, for example 70% of those at prose Level 1. Fourteen per cent of those at Level 1 on the prose scale had only received education to primary level. This compares with only 29% of those at Level 4/5 who had finished their education at lower secondary level and 2% who had only received primary education.

  (Figure 5.2, Table A5.2)

- A small proportion (3% on the prose scale) of those performing at Level 1 had received higher education.

  (Figure 5.2, Table A5.2)

- Compared to those with high literacy skills people with low literacy skills were more likely to be unemployed (13% on the document scale compared with 6%) or economically inactive (36% compared with 5% (excluding students)). It was also the case, however, that large numbers of people with poor skills were employed; half of those who performed at Level 1 on each of the three scales were employed.

  (Figure 5.3, Table A5.3)

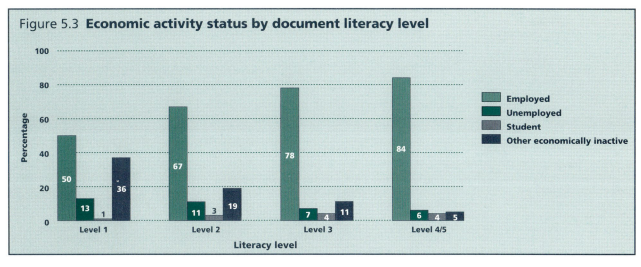

Figure 5.3 **Economic activity status by document literacy level**

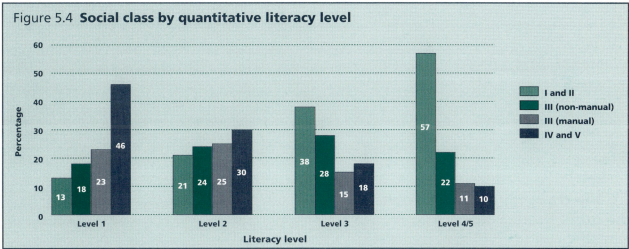

Figure 5.4 **Social class by quantitative literacy level**

- Those at Level 1 were most likely to belong to the semi-skilled and unskilled manual social groups (46% on the quantitative scale).

  (Figure 5.4, Table A5.4)

- People performing at Level 1 were more likely than those with high skills to have a low personal income (in the lowest two quintiles, less than £5,929 per year); 63% of those at Level 1 on the document scale were in this category.

  (Table 5.2)

- Over a third (37%) of people at Level 1 on the prose scale received social security benefits (excluding pensions and child benefit) compared with only 9% of those with higher literacy skills.

  (Figure 5.5, Table A5.5)

- Around 14% of those at Level 1 did not speak English as their first language as a child which compares with around 3% of those with a higher level of literacy skills.

  (Table 5.3)

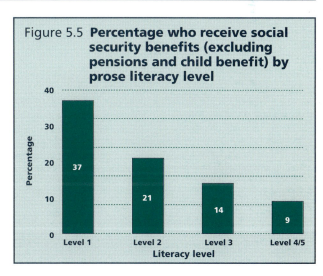

Figure 5.5 **Percentage who receive social security benefits (excluding pensions and child benefit) by prose literacy level**

- 15% of those at Level 1 on the prose scale were not born in the UK and 14% were in a non-white ethnic group.

  (Table 5.3)

Table 5.2 Gross personal income by literacy level

	Quintile 1-2** (less than £5,929 per year)	Quintile 3 - 5** (£5,929 or more per year)	Total	Base*
	%	%	%	
Prose literacy				
Level 1	60	40	100	774
Level 2	43	57	100	1115
Level 3	34	66	100	1163
Levels 4/5	28	72	100	606
Total	41	59	100	3658
Document literacy				
Level 1	63	37	100	843
Level 2	44	56	100	1016
Level 3	34	66	100	1114
Levels 4/5	23	77	100	685
Total	41	59	100	3658
Quantitative literacy				
Level 1	63	37	100	851
Level 2	46	54	100	1025
Level 3	34	66	100	1106
Levels 4/5	18	82	100	676
Total	41	59	100	3658

*Informants who refused or did not know their income were not included in the analysis.
** The quintile bands were calculated from the personal income data of individuals aged 16-65 from the General Household Survey.

Table 5.3 Percentage of respondents who a) first spoke a language other than English, b) were not born in the UK and c) belong to a non-white ethnic group by literacy level

	First spoke language other than English	Not born in UK	Non-white ethnic group	Base
	Percentage of respondents			
Prose literacy				
Level 1	14	15	14	825
Level 2	7	6	6	1165
Level 3	3	6	3	1204
Levels 4/5	2	6	2	617
Total	6	8	6	3811
Document literacy				
Level 1	13	15	13	900
Level 2	6	6	5	1057
Level 3	4	5	4	1154
Levels 4/5	3	6	2	700
Total	6	8	6	3811
Quantitative literacy				
Level 1	14	14	13	905
Level 2	4	7	5	1070
Level 3	4	5	4	1146
Levels 4/5	3	6	3	690
Total	6	8	6	3811

5.3 The characteristics of people at the other literacy levels

The previous section showed that there is a clearly defined group of people performing at Level 1 on the literacy scales. This section will show that there is also a clearly defined group of people performing at Level 4/5 while those performing at Levels 2 and 3 do not fall into homogenous groups.

5.3.1 People at Level 2

People performing at Level 2 can deal with more complex and more varied texts than people at Level 1. They can locate or integrate one or more pieces of information in a text where there may be distracting information and where low level inferences may be required. Those performing at Level 2 were a less homogenous group than those at Level 1. In many respects their profile was similar to that of the wider population; they were fairly evenly split across the age-groups, across manual and non-manual social classes and between the low and medium-high income quintiles. The majority had continued in education until lower secondary level.

- Those at Level 2 on the document scale were more likely to be women than men but there was no significant difference between the sexes on the other two literacy scales.

  (Table 5.1)

- 65% of those at Level 2 on the prose scale had continued in education until lower secondary level which is slightly higher than the proportion in the general population while the proportion who had gone into further education was the same as that in the general population (20%).

  (Figure 5.2, Table A5.2)

- A higher proportion of those at Level 2 than at Level 1 were in employment; 67% on the document scale compared with 50%.

  (Figure 5.3, Table A5.3)

- Around one fifth of those at Level 2 received social security benefits (excluding pensions and child benefit) which is a lower proportion than at Level 1 but higher than Level 4/5.

  (Figure 5.5, Table A5.5)

5.3.2 People at Level 3

Level 3 tasks involve more complex operations such as matching information that requires low level inferences or drawing out information from different parts of a text. They involve more complex and varied mathematical operations such as multiplication and division where the appropriate operation must be inferred from the text. People at Level 3 tended to be younger and better educated than those at Levels 1 and 2. The majority were employed, mostly in non-manual occupations.

- Around half had received education to lower secondary level, a quarter had gone onto further education and the rest had progressed to higher education. The proportion continuing into higher education was greater than that for Levels 1 and 2 but lower than that for Level 4/5; 26% compared with 43% of those at Level 4/5 on the prose scale.

(Figure 5.2, Table A5.2)

- The majority (66%) had a personal gross income in Quintiles 3-5 and, on the prose scale, only 14% received social security benefits (excluding pensions and child benefit).

(Tables 5.2 and A5.5, Figure 5.5)

5.3.3 People at Level 4/5

People at this level can handle a wide variety of lengthy, complex and dense texts that require the reader to integrate several pieces of information and make inferences based on the text. They are able to solve mathematical problems where the operation required is not explicitly stated and which may require background knowledge. They have trouble only with some of the most difficult items. Those at Level 4/5 were predominantly young, with high levels of education, although a notable minority (29% on the prose scale) had not continued their formal education beyond lower secondary level. On the document and quantitative scales they were more likely to be men. People at Level 4/5 were the most likely to be in employment and to be in non-manual social classes.

- On the prose scale 24% were aged 16-25, 27% were aged 26-35 and 28% were aged 36-45 which meant that over three quarters were aged 45 or under.

(Figure 5.1, Table A5.1)

- 69% of those at Level 4/5 on the quantitative scale were men and 31% were women.

(Table 5.1)

- 43% of those at Level 4/5 on the prose scale had received higher education.

(Figure 5.2, Table A5.2)

- 30% of those at Level 4/5 on the prose scale had not continued their education beyond lower secondary level showing that it is possible to develop high literacy skills outside the context of formal education.

(Figure 5.2, Table A5.2)

- Over half were in Social Classes I and II, around a fifth were in a manual social class.

(Figure 5.4, Tables A5.4)

- The proportion with a personal gross income in the upper three quintiles was high; 82% on the quantitative scale.

(Table 5.2)

- Few (9%) of those at Level 4/5 on the prose scale said that they received social security benefits (excluding pensions and child benefit).

(Figure 5.5, Table A5.5)

5.4 Literacy practice of people with low skills

In addition to education and employment, the other means of improving and maintaining literacy skills is by literacy practice at home and in the community. Literacy practice in daily life has been discussed in Chapter 4. This section will particularly look at whether people with low literacy skills differ in this type of behaviour from those with higher skills.

The report on the first round of IALS[2] found that those who watch television for more than five hours a day usually have lower levels of literacy. A quarter of those at Level 1 on the prose scale reported watching television for five hours or more per day compared with only 4% of those at Level 4/5. This does not imply that watching television causes people to have low literacy skills. As shown earlier in the chapter those at Level 4/5 were more likely to be younger and in employment than those at Level 1 and would therefore

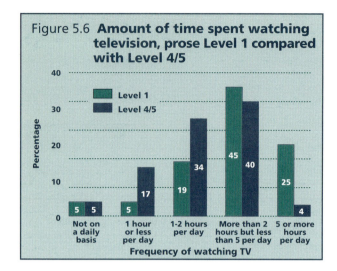

Figure 5.6 **Amount of time spent watching television, prose Level 1 compared with Level 4/5**

have less opportunity to watch television whereas those at Level 1 were more likely to be older, unemployed and on a low income and so may have more free time available to watch television.

(Figure 5.6, Table A5.6)

Respondents were asked how often they read books outside of work; on each of the scales those at Level 1 were less likely than those at Level 4/5 to say they read a book daily (16% compared with 46% on the prose scale) or that they read a book weekly (10% compared with 23% on the prose scale). As shown in Chapter 4 this may be related to the amount of time people at Level 1 spent watching television; people who watched television for five hours or more per day were not as likely as those who spent less time watching television to read a book at least once a week.

(Figure 5.7, Table A5.7)

The majority of people at all levels of literacy said that they read a newspaper or magazine at least once a week.

However, there is a wide variety of newspapers and magazines available. Although people were not asked to specify which newspapers they read, they were asked to say which parts of a newspaper they generally read. People with high literacy skills were more likely than those with low skills to read sections of the newspaper such as national/international news, the editorial page, financial news and book, film or art reviews. The only two sections which people at Level 1 were more likely than those at Levels 4/5 to read were the classified advertisements (43% compared with 33% on the prose scale) and the horoscopes (46% compared with 31% on the prose scale). This may reflect the types of newspapers that people at Level 1 read compared with the papers read by those at Level 4/5. Those with higher literacy skills may be more likely to read the broadsheets which do not contain horoscopes. People at Level 1 may be more likely to read the classified advertisements as they are more likely to be unemployed and may be reading the job adverts.

(Tables A5.8 and A5.9)

Respondents were also asked whether they followed current events. Over half of those with low quantitative literacy skills said that they followed current affairs most or some of the time but 19% said that they hardly followed current events at all and 24% said that they only followed them now and then. Of people at Levels 4/5, 94% said that they followed current affairs most or some of the time and only 1% said that they hardly followed them at all. Compared with those at all the other literacy levels those at Level 1 appear the least interested in current events.

(Figure 5.8, Table A5.10)

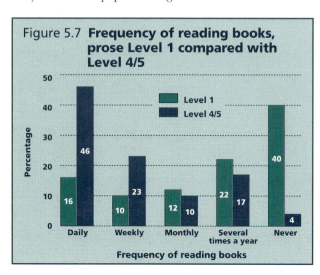

Figure 5.7 **Frequency of reading books, prose Level 1 compared with Level 4/5**

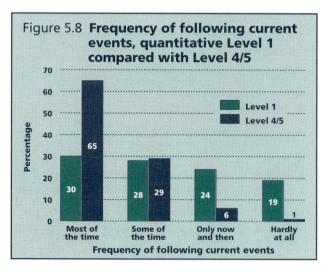

Figure 5.8 **Frequency of following current events, quantitative Level 1 compared with Level 4/5**

5.5 Factors affecting the probability of being at Level 1

We have seen that the low skilled form a fairly homogenous group. However, the characteristics and behaviours of the low-skilled group interrelate, for example older people are more likely to have low levels of literacy skill than younger people but this may be because they tend to have lower levels of education than younger people. Education may be a more important factor in determining their literacy level than age but the analysis we have carried out so far, looking at each factor separately, does not tell us whether this is the case or not. Logistic regression (a multivariate statistical technique) can be used to look at how factors interact with each other.

A separate logistic regression was carried out on each of the three literacy scales to look at the odds or probabilities[3] of people being at Level 1. The following characteristics and behavioural practices were included in the analysis:

- education level;
- age;
- sex;
- social class;
- gross personal income;
- receipt of benefits;
- language first spoken as a child;
- ethnic group;
- time spent watching television;
- frequency of reading books.

For each of the variables included in the regression, a coefficient is produced which represents the factor by which the odds of a person being at Level 1 on each of the scales increases if a person has that characteristic. One of the categories of the variable is defined as the reference category (with a value of 1.00) and the odds are given in comparison to this. For example, the reference category for education level is higher education; if the values of all the above variables are the same except for education level then Table 5.4 shows that those with the lowest level of education are 5.8 times more likely to be at Level 1 on the prose scale than those with higher education.

The odds of being at Level 1 for a combination of characteristics can be calculated by multiplying the baseline odds shown at the bottom of the table by the appropriate factors. So, for example, a person educated to lower secondary level or below, aged 45 or over, in Social Class III (manual), with a personal income of £2,705-£5,928 per year, not in receipt of benefits, who is white, watches more than five hours of television per day and reads a book less often than once a week would have odds calculated as shown below for the prose scale:

$$0.0043 \times 5.80 \times 2.85 \times 2.05 \times 3.07 \times 1.00 \times 1.00 \times 1.73 \times 2.70 = 2.01$$

This would give odds of 2.01 to 1 that a person with these characteristics would be at Level 1 on the prose scale. The odds can be converted to a probability using the following formula:

$$p = \frac{odds}{1 + odds}$$

so the probability would be 67%. The model only shows the likelihood of people with given characteristics being at Level 1; it does not imply that all people with such characteristics will be at Level 1 on the literacy scales nor does it imply that these characteristics necessarily cause low literacy.

Table 5.4 shows that level of education was the most important factor on all three scales in predicting whether or not a person would be at Level 1. Holding all other factors constant the odds of a person with the lowest level of education being at Level 1 were at least four times the odds of a person with the highest level of education being at this level. Age is also an important factor, although there are differences between the scales. The odds of a person aged 45 or over being at Level 1 on the quantitative scale are 1.76 times that of a person aged 16-24 while on the prose scale the odds are 2.85. Social class and income were also important factors.

(Table 5.4)

Table 5.4 Odds of an individual being at Level 1 on the prose, document and quantitative literacy scales

Characteristics	Prose		Document		Quantitative	
	Multiplying factors	95% confidence intervals	Multiplying factors	95% confidence intervals	Multiplying factors	95% confidence intervals
Education level						
Second level, 1st stage or lower	5.80 *	3.74-9.02	4.06 *	2.78-5.92	4.33 *	2.95-6.36
Second level, 2nd stage	2.73 *	1.68-4.45	1.86 *	1.21-2.86	1.97 *	1.27-3.04
Third level	1.00		1.00		1.00	
Age-group						
16-24	1.00		1.00		1.00	
25-44	1.30	0.93-1.82	1.28	0.93-1.78	1.01	0.74-1.38
45 or over	2.85 *	2.04-3.98	2.73 *	1.98-3.77	1.76 *	1.30-2.39
Sex**						
Male	-		1.00		1.00	
Female	-		1.31 *	1.04-1.64	1.64 *	1.31-2.06
Social class						
I and II	1.00		1.00		1.00	
III (non-manual)	0.75	0.53-1.06	0.80	0.58-1.11	0.98	0.71-1.36
III (manual)	2.05 *	1.48-2.83	2.28 *	1.65-3.15	2.38 *	1.71-3.30
IV and V	2.49 *	1.83-3.41	2.32 *	1.72-3.14	2.66 *	1.96-3.62
Gross personal income						
Up to £2,704 p.a.	2.89 *	1.85-4.50	3.35 *	2.13-5.26	2.92 *	1.85-4.60
£2,705 - £5,928 p.a.	3.07 *	1.98-4.78	3.96 *	2.53-6.21	2.91 *	1.85-4.59
£5,929 - £10,400 p.a	2.78 *	1.81-4.25	2.83 *	1.84-4.36	2.70 *	1.75-4.18
£10,401 - £16,848 p.a	1.55	0.99-2.43	1.68 *	1.08-2.63	1.50	0.95-2.36
£16,849 or more p.a.	1.00		1.00		1.00	
Receipt of benefits						
Receives benefits	1.41 *	1.12-1.78	1.55 *	1.24-1.93	1.68 *	1.35-2.09
Does not receive benefits	1.00		1.00		1.00	
Language first spoken						
English	1.00		1.00		1.00	
Not English	2.55 *	1.67-3.90	2.33 *	1.54-3.53	2.79 *	1.85-4.20
Ethnic group						
White	1.00		1.00		1.00	
Not white	2.84 *	1.67-4.82	2.50 *	1.49-4.20	1.88 *	1.12-3.15
Time spent watching television						
Less than 5 hours per day	1.00		1.00		1.00	
5 or more hours per day	1.73 *	1.35-2.22	1.71 *	1.34-2.19	1.79 *	1.41-2.28
Frequency of reading books						
Less than weekly	2.70 *	2.20-3.32	2.38 *	1.95-2.90	1.93 *	1.59-2.34
At least weekly	1.00		1.00		1.00	
Number of cases in the model	3430		3430		3430	
Baseline odds	0.0043		0.0058		0.0075	

* Significant at the 95% level
** Sex was not included as one of the variables in the model for prose literacy

5.6 Self-assessment of literacy skills

Although the proportion of people with low skills may be a problem for an economy which is increasingly demanding a workforce with higher literacy skills (see Chapter 3, Section 3.2) the important question in addressing this problem may be whether individuals themselves recognise that they have low levels of literacy and whether or not they see it as a problem. People are unlikely to address their skills deficit if they do not recognise it as a problem. Respondents were asked to rate their reading, writing and mathematics skills for both their job and for everyday life outside of the work context. Among those who performed at the lowest literacy level a large proportion considered their skills to be adequate for daily life; around three quarters were 'very satisfied' or 'somewhat satisfied' with their reading and writing skills. Women were more likely than men to describe themselves as 'very satisfied' with their reading and writing skill whereas men were more likely than women to describe themselves as 'very dissatisfied'; 10% of men at prose Level 1 described themselves as 'very dissatisfied' compared with 5% of women.

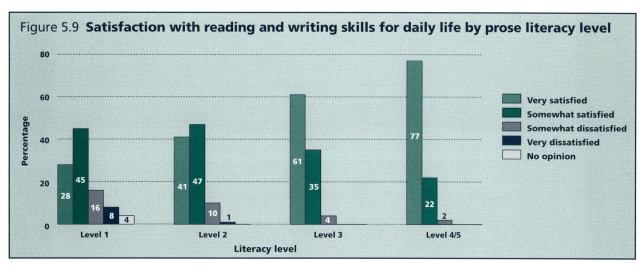

Figure 5.9 **Satisfaction with reading and writing skills for daily life by prose literacy level**

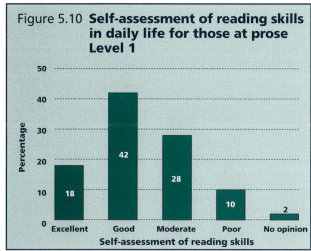

Figure 5.10 **Self-assessment of reading skills in daily life for those at prose Level 1**

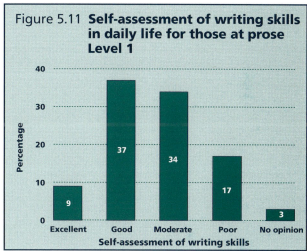

Figure 5.11 **Self-assessment of writing skills in daily life for those at prose Level 1**

Younger people were more likely than older people to describe themselves as 'very satisfied' (Table not shown). The proportion of people at Level 2 who were 'very satisfied' or 'somewhat satisfied' with their reading and writings skills was also high (41% and 47% on the prose scale).

(Figure 5.9, Table A5.11)

Although people with low skills were least likely to rate their reading skills for everyday life as excellent, there were, nevertheless, 18% of those at prose Level 1 who rated their reading skills as excellent and a further 42% who rated their skills as good. Women were more likely than men to say their reading skills were excellent or good; 20% and 44% respectively compared with 16% and 40%. Only one in ten people at prose Level 1 rated their reading skills as poor.

(Figure 5.10, Table A5.12)

A lower proportion (9%) of people at Level 1 on the prose scale rated their writing skills as excellent and 37% rated them as good; 17% rated their writing skills as poor. Again women were more likely than men to say their writing skills were excellent or good and less likely to describe them as poor. A similar pattern was shown for people at Level 1 and Level 2 on the document scale and for the questions relating to reading and writing skills for work.

(Figure 5.11, Tables A5.13)

Looking at quantitative literacy people were less likely to rate their maths skills as highly as they had done for reading or writing. Of those performing at Level 1 on quantitative literacy, 18% recognized that their maths skills were poor and a further 45% rated their mathematical ability as moderate. A small proportion of respondents at Level 1 (5%) perceived their mathematical skills as excellent and 29% said they were good. At Level 1 there was little difference between men and women in their self-assessment of

61

maths skills. People may be more likely to admit to having problems with writing and maths than they are to admit to having trouble reading as it might be more socially acceptable to have problems with writing and maths than it is to have problems with reading .

(Figure 5.12, Table A5.14)

People with low literacy skills may adapt their lives to avoid situations where they need to use their literacy skills and therefore remain satisfied with their skill level. However, when they do need to carry out tasks which require a higher level of literacy they may then have to ask others for help. The survey asked people if they ever needed help with the following:

- reading newspaper articles;
- reading information from government departments, businesses or other institutions;
- filling out forms such as applications or bank deposit slips;

- reading instructions such as on a medicine bottle;
- reading instructions on packaged goods in shops or supermarkets;
- doing basic arithmetic, that is, adding, subtracting, multiplying and dividing;
- writing notes and letters.

Although people at Level 1 were on the whole satisfied with their literacy skills they were more likely than those with higher skills to say that they sometimes or often needed help with each of the above tasks, suggesting that they were in some sense aware of their limitations. The tasks that people with low skills particularly reported needing help with were filling out forms and reading information from government departments, businesses and other institutions; on the prose scale 46% of those at Level 1 and 22% of those at Level 2 said they needed help filling out forms. Forty per cent of those at Level 1 and 26% of those at Level 2 said that they sometimes or often needed help reading information from government departments, businesses or other institutions. This is particularly important given the findings reported earlier that people at Level 1 were more likely than those at higher literacy levels to be unemployed or economically inactive and more likely to be receiving social security benefits.

(Figure 5.13, Table A5.15,A5.16)

Twenty nine per cent of those at Level 1 on the quantitative scale said that they sometimes or often needed help doing basic arithmetic. Thirty per cent of those at Level 1 on the prose scale said that they

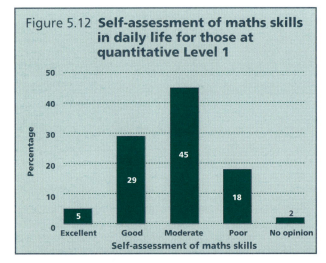

Figure 5.12 **Self-assessment of maths skills in daily life for those at quantitative Level 1**

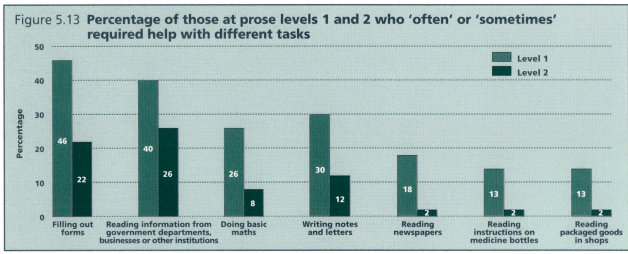

Figure 5.13 **Percentage of those at prose levels 1 and 2 who 'often' or 'sometimes' required help with different tasks**

needed help writing notes and letters as did 12% of those at Level 2.

(Figure 5.13, Table A5.17, A5.18)

Eighteen per cent of those at Level 1 on the prose scale said that they often or sometimes needed help reading a newspaper whereas only 2% of those at Level 2 said that they needed help. There was a similar pattern among those at Level 1 on the document and quantitative scales.

(Figure 5.13, Table A5.19)

Thirteen per cent of people at Level 1 sometimes or often needed help reading instructions on a medicine bottle. A similar proportion said they needed help reading instructions on packaged goods in shops.

(Figure 5.13, Tables A5.20, A5.21)

Notes and references

1. Bynner John and Parsons Samantha. *It doesn't get any better The impact of poor basic skills on the lives of 37 year olds.* The Basic Skills Agency (London 1997). One of the findings reported is that people with low literacy skills were less likely than those with higher skills to be employed and if they were employed they were more likely to work in casual or unskilled jobs.
2. *Literacy, Economy and Society Results of the first International Adult Literacy Survey* OECD and Statistics Canada (1995).
3. Logistic regression can be used to predict the odds of behaviour occurring for different combinations of independent variables (odds refers to the ratio of the probability that the event will occur to the probability that the event will not occur).

6 Literacy skills in other countries

6.1 Introduction

The British survey formed part of an international programme of surveys carried out in a number of countries worldwide. The report on the seven countries that took part in the first round of the study was published in 1995.[1] A further volume, due for publication in Autumn 1997,[2] will include additional countries, including the United Kingdom, that have since participated in the study. The preceding chapters have described the results of the British Adult Literacy survey with little reference to the relative performance of the other countries that took part. This chapter discusses the British results in the international context examining how literacy skills are distributed in the different countries across the main socio-demographic characteristics.

6.2 Comparability of data

There is increasing interest in making international comparisons as the process of globalisation advances. International organisations are seeking more and more to standardise and harmonise statistics across countries.[3] There is particular interest in the relative performance of different countries in terms of establishing or maintaining competitive advantage. However, in pursuing this interest the issue of comparability is sometimes overlooked. While every effort was made in IALS to produce statistics that were comparable, there are differences between the surveys (for example in response rates and sample design) and these should be borne in mind when comparing the relative performance of different countries. (See also Chapter 1, section 1.4)

One measure of the reliability of a survey estimate is the standard error[4], and these are shown in tables where available. However, standard errors reflect only the error that comes from sampling (the sample design and the sample size), taking into account the variability of the characteristic being measured, in this case literacy. They do not take account of non-sampling error or of differences between countries in sample design and implementation and are likely, therefore, to underestimate the total error.

For the purposes of international comparisons each country had to supply data in a prescribed format and structure. The classifications used for some variables differed from those that would normally be used in this country, for example the classification of occupations and industry and economic activity status. In the preceding chapters the survey results are reported using the standard classifications that apply in Britain. In this chapter the international rubrics are used and so the tables may differ slightly from those presented in the earlier chapters.[5] Data for all the countries are not always shown in each figure or table. Where only a selection of countries is shown the data for all countries are shown in the relevant table in the annex.

6.3 How literacy skills are distributed

The distribution of prose, document and quantitative literacy skills differ between countries. As Figure 6.1 shows, some countries, such as Sweden and Poland, had very skewed distributions across the different skill levels with the majority of the population at either the lower or upper end of the distribution. Other countries, such as Germany and the Netherlands, had more centrally distributed skill levels with only small proportions performing at the upper or lower levels and the majority of the population performing in the middle skill levels. Relative to other countries, the United States and Britain had a more uniform distribution across the different skill levels and although the majority of the population were at the middle skill levels, large proportions of the population were also at the lowest or highest literacy levels.

Britain's distribution is similar to that in the other English speaking countries, the United States and Canada, with slightly higher proportions at Levels 2 and 3 than there are at Level 1 and Level 4/5. In Britain, just over a fifth (22%) of the population of working age performed at Level 1 on the prose scale while 30% and 31% performed at Levels 2 and 3 respectively and 17% performed at the two highest levels combined (Level 4/5). In Germany, the Netherlands and Switzerland the majority of the population of working age performed at literacy

Levels 2 and 3 with smaller proportions at Levels 1 and 4/5. Sweden and Poland had the most skewed distributions, a large proportion of the Swedish population performed at the higher skill levels and very few were at Level 1 while in Poland the majority were at the lower skill levels and very few were at Level 4/5. When comparing proportions at the upper and lower literacy levels across countries, Britain, the US and Canada, with literacy skills distributed more uniformly across the different literacy levels, appear more polarised compared with other countries, as they have relatively large proportions of the population at both the lower and upper literacy levels.

(Figure 6.1, Table 6.1)

To illustrate the importance of taking account of sampling error, Figure 6.2 shows the confidence intervals around the estimates of the proportion of the population at the upper and lower ends of the skill distribution in each country (Level 1 and Level 4/5). If one hundred separate random samples were taken, the estimate of those at Level 1 or those at Level 4/5 in each country would fall within the ranges shown in ninety five of those samples. As already stated in section 6.2, they are likely to underestimate the total error as they do not take account of non sampling error. In terms of statistical significance, where there is an overlap in the confidence intervals the differences between the estimates for those countries will not be significant. So for example, although a higher proportion of Britons were at Level 1 on prose compared with all countries except Poland, there were only three countries - Sweden, the Netherlands and Germany - which had significantly lower proportions than Britain at Level 1, as the range of estimates for other countries overlap. The proportion of Britons at prose Level 1 was not significantly different from the proportion at that level in the US, Canada or either of the two language groups in Switzerland. The relative position of Britain to other countries differs when considering the proportion of the population at

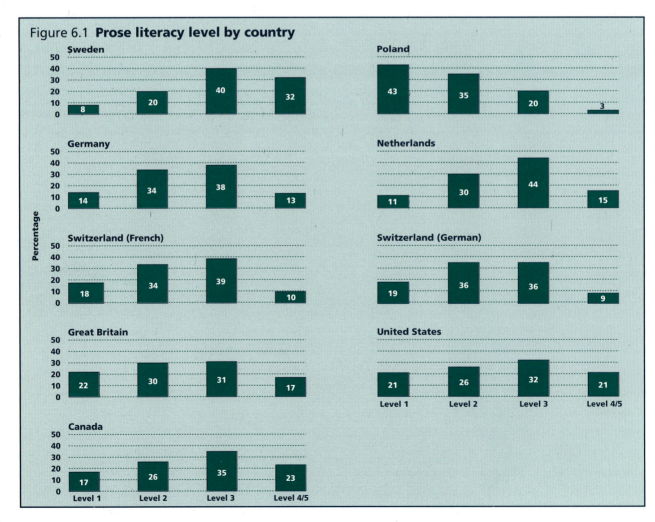

Figure 6.1 **Prose literacy level by country**

Table 6.1   Literacy level by country

	Level 1		Level 2		Level 3		Level 4/5		Total
	%	s.e.	%	s.e.	%	s.e.	%	s.e.	%
Prose literacy									
Canada	17	1.6	26	1.8	35	2.4	23	2.3	100
Germany	14	0.9	34	1.0	38	1.3	13	1.0	100
Great Britain	22	1.0	30	1.3	31	1.2	17	0.8	100
Netherlands	11	0.6	30	0.9	44	1.0	15	0.6	100
Poland	43	0.9	35	0.9	20	0.7	3	0.3	100
Sweden	8	0.5	20	0.6	40	0.9	32	0.5	100
Switzerland (French)	18	1.3	34	1.6	39	1.8	10	0.7	100
Switzerland (German)	19	1.0	36	1.6	36	1.3	9	1.0	100
United States	21	0.7	26	1.1	32	1.2	21	1.2	100
Document literacy									
Canada	18	1.9	25	1.5	32	1.8	25	1.3	100
Germany	9	0.7	33	1.2	40	1.0	19	1.0	100
Great Britain	23	1.0	27	1.0	31	1.0	19	1.0	100
Netherlands	10	0.7	26	0.8	44	0.9	20	0.8	100
Poland	45	1.3	31	1.0	18	0.7	6	0.3	100
Sweden	6	0.4	19	0.7	39	0.8	36	0.6	100
Switzerland (French)	16	1.3	29	1.4	39	1.3	16	1.1	100
Switzerland (German)	18	1.0	29	1.5	37	0.8	16	1.0	100
United States	24	0.8	26	1.1	31	0.9	19	1.0	100
Quantitative literacy									
Canada	17	1.8	26	2.5	35	2.1	22	1.8	100
Germany	7	0.4	27	1.2	43	0.8	24	0.9	100
Great Britain	23	0.9	28	1.0	30	0.9	19	1.0	100
Netherlands	10	0.7	26	0.9	44	1.0	20	0.8	100
Poland	39	1.1	30	1.2	24	0.6	7	0.5	100
Sweden	7	0.4	19	0.6	39	0.9	36	0.7	100
Switzerland (French)	13	0.9	25	1.4	42	1.6	20	1.0	100
Switzerland (German)	14	1.0	26	1.3	41	1.5	19	1.3	100
United States	21	0.7	25	1.1	31	0.8	23	1.0	100

s.e. Standard error of the estimate. The reported sample estimate can be said to be within 2 standard errors of the true population value with 95% confidence.

the highest prose literacy levels, rather than the lowest, only Sweden and the US had significantly higher proportions at Level 4/5 on prose compared with Britain; 32% of Swedes and 21% of the US population performed at Level 4/5 on prose compared with 17% of the British population. There was no significant difference between Britain and the Netherlands, Germany or Canada in the proportion who were at the highest level on prose literacy.

(Figure 6.2)

On quantitative literacy, only Sweden and Germany had a significantly higher proportion of their population than Britain at Level 4/5. At the lower literacy level Britain had a significantly higher proportion at Level 1 on quantitative literacy than most other countries except the US. Almost a quarter (23%) of Britons of working age were at Level 1 on quantitative literacy compared with just 10% or less in the Netherlands, Germany and Sweden. In a recent

study of 13 year olds however Sweden, Germany and England had very similar scores on mathematics.[6]

(Table 6.1)

Some countries showed greater consistency between the prose, document and quantitative literacy scales than others. In the Netherlands, Great Britain and Canada there was little difference between the scales in the proportion performing at each level. In Germany and in the two language groups in Switzerland the distribution was different on each scale with fewer performing at the higher levels on prose than on the other two scales. In Germany for example, 13% performed at Level 4/5 on prose while 19% and 24% performed at that level on document and quantitative literacy respectively.

(Table 6.1)

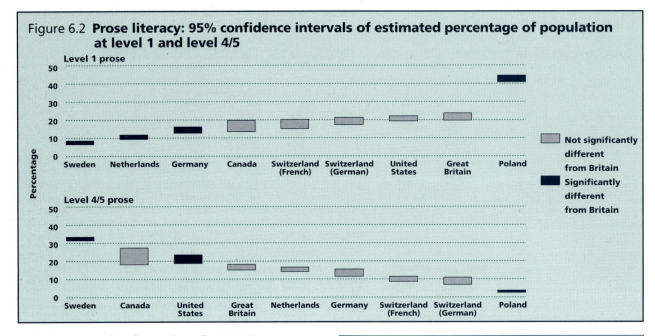

Figure 6.2 **Prose literacy: 95% confidence intervals of estimated percentage of population at level 1 and level 4/5**

6.4 Age and education in IALS countries

In earlier chapters we have seen that both age and educational attainment level are related to performance on the literacy assessment. How countries perform overall will partly be a function of how these characteristics and other related characteristics are distributed in the population. The distribution of the population in terms of age and educational level is not the same in each country as the following table and figures show. For example, Germany has a higher percentage of its population in the oldest age-group than other countries while the US distribution tends towards the younger end of the age range. If all other things were equal this would result in Germany having more people at the lower literacy levels and the US fewer, just because of their age distribution.

(Table 6.2)

The educational attainment level of the population of working age is changing fairly rapidly as older people, who have on average lower levels of education, move out of the work force. Table 6.3 shows the educational attainment level of the population aged 25[7] to 64 for some countries that took part in IALS. The Netherlands has a much higher proportion educated to second level, 1st stage (ISCED 2) than the other countries. Canada and the United States have higher proportions of their population who

Table 6.2 **Population aged 15 to 64: Age-group by country**

| Country | Age-group | | | | | |
	15-24	25-34	35-44	45-54	55-64	Total
	%	%	%	%	%	
Canada	20	25	25	18	13	100
Germany	17	25	21	18	18	100
Netherlands	20	25	22	19	14	100
Sweden	20	22	21	22	15	100
Great Britain	19	25	21	20	15	100
US	21	24	25	18	12	100

Source: Germany, Netherlands and Sweden, Demographic Statistics 1996 Eurostat Table B13 page 29. Population by age group on 1 January 1995 Figures for Canada are for 1 July 1995. Statistics Canada. Annual Demographic Statistics 1996. Table 1.4 page 70
Figures for US from Statistical Abstract of the United States 1996 The National Data Book. US Department of Commerce, Economics and Statistical Administration, Bureau of the Census. October 1996 table 14 page 15
Figures for Great Britain are mid-1995 estimates from the Population Estimates Unit, ONS.

have completed third level education; almost one in four Americans (24%) aged 25-64 have completed a university programme compared with just 12% of the United Kingdom population in the same age range and 13% of Germans.

(Table 6.3)

These differences are not simply due to the different age profiles of the countries, there are also differences in educational attainment within age-groups or cohorts. Table 6.4 shows the percentage who have attained at least upper secondary education (ISCED 3) in each country for different age-groups. Among the older age-groups, 84% of those aged 45-54 in Germany and 85% of the same age-group in the United States had completed at least upper secondary

Table 6.3 Highest level of educational attainment (ISCED) of population aged 25 to 64 by country (1994)

	Highest level of educational attainment				
	Second level, 1st stage or lower	Second level, 2nd stage	Third level, non-university	Third level, university	Total
	%	%	%	%	%
Canada	26	28	29	17	100
United States	15	53	8	24	100
Great Britain	25	54	9	12	100
Sweden	28	46	14	12	100
Switzerland	18	61	13	8	100
Netherlands	40	38	..	21	100
Germany	16	62	10	13	100

Source: Education at a glance OECD; Paris 1996 Table C1 page 35
Figures for Great Britain supplied by the Scottish Office
Under 25s excluded as a large proportion have not yet completed their education

Table 6.4 Percentage who had attained at least upper secondary education (ISCED 3) by age-group and country (1994)

Country	Age-group			
	25-34	35-44	45-54	55-64
	Percentage who had attained ISCED 3			
Canada	82	79	70	53
United States	86	89	85	76
Great Britain	86	79	69	58
Sweden	85	78	69	52
Switzerland	89	84	79	73
Netherlands	69	64	54	44
Germany	90	88	84	72

Source: Education at a glance OECD (1996 Paris) Table c1.2 page 36
Figures for Great Britain supplied by The Scottish Office

education compared with about 70% of the Canadian, Swedish and British population of the same age and 54% of the Netherlands population. Apart from the Netherlands the proportion of those aged 25-34 who attained at least upper secondary education in each country is fairly similar ranging from 82% of Canadians to 90% of Germans in that age-group.

(Table 6.4)

6.5 Literacy skills and age

In all countries except one, the proportion of the population at literacy Levels 1 and 2 on each of the three literacy dimensions is highest in the oldest age-groups. The exception is the United States where there is also a high proportion of young people at these levels. In Britain, in all but the oldest age-group the distribution of literacy skills is very similar. The same is true in Sweden, although in all age-groups the proportion of Swedes who perform at the two lowest literacy levels is much lower than in other countries. Given the increase in participation in post-16 education and increasing rates of educational qualifications in Britain the performance of the two youngest age-groups might be viewed as somewhat disappointing.

(Figure 6.3, Table A6.1)

The importance of bearing in mind the standard error around the literacy estimates when making comparisons between countries has already been discussed. This is equally important when considering differences between countries in the literacy skills of the different age-groups, as shown in Figure 6.4. For example, while the proportion in each age-group in Britain at prose Level 1 appears to show a relatively poor performance compared with other countries, the differences are often not statistically significant. Among the youngest age-group there was no significant difference in the proportion at prose level 1 between Great Britain, the US, Canada or among the French speaking group in Switzerland. Among those aged 26-35 only Sweden, the Netherlands and the French speaking group in Switzerland had significantly lower proportions at prose Level 1 than Britain. Among those aged 36-45 and 46-55 the differences between countries were even less clearly defined with only Sweden and the Netherlands having significantly lower proportions at Level 1 than Britain. Among most age-groups there was no significant difference between Britain and other countries in the proportion performing at Level 2.

(Figure 6.4, Table A6.1)

There was less difference in the relative performance of countries in the proportion of each age-group who were at the highest prose level (Level 4/5). Among those in the 16-25 and 26-35 age-groups there was no significant difference between Britain and most countries. In the youngest age-group only Sweden and Poland were significantly different from Britain in the proportion who performed at Level 4/5 on prose; 40% of Swedes aged 16-25 performed at Level

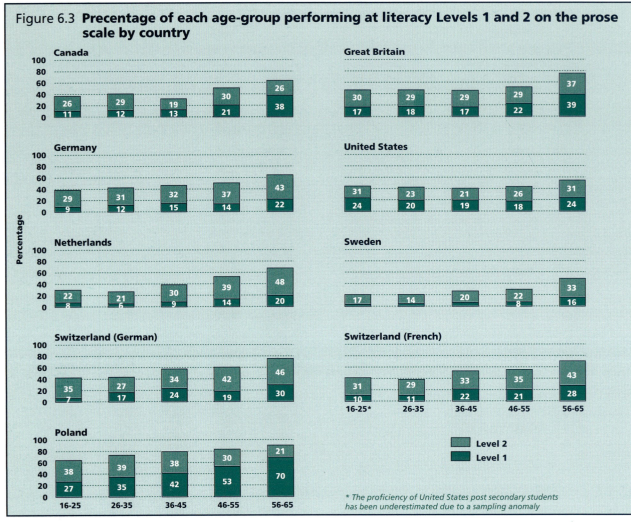

Figure 6.3 **Precentage of each age-group performing at literacy Levels 1 and 2 on the prose scale by country**

* The proficiency of United States post secondary students has been underestimated due to a sampling anomaly

4/5 on prose compared with 20% of Britons in that age-group and 6% of Poles. A similar pattern was evident among those aged 26-35 with only Sweden, Poland and the German speaking group in Switzerland having significantly different proportions at that Level compared with Britain.

(Figure 6.5, Table A6.1)

6.6 Literacy skills and educational attainment

In all countries those with more education tended to perform at higher literacy levels on all three scales although the relationship is stronger in some countries than in others. This section discusses the pattern in relation to prose literacy but the findings are similar for all three dimensions. There is a broad incremental shift in the modal literacy level of the population at the different education levels, that is,

the level where the majority are located. With the exception of Sweden and the Netherlands those with the lowest level of educational attainment are predominantly at prose Level 1. In most countries those with lower secondary education (ISCED 2) are predominantly at Level 2 although some countries have considerable proportions with this level of education at either Level 1 (Britain and the US) or Level 3 (the Netherlands and Sweden).

At the next level of education, upper secondary level (ISCED 3), the majority of people perform at Level 2 or 3 on prose, again with the exception of Sweden where a large proportion of those with this level of education are at Level 4/5. In all countries except Poland, those with higher education are predominantly at Level 3 and 4/5. In Canada and Sweden those with third level, university education were predominantly at Level 4/5 while in the US, Germany and Britain those with this level of education were more evenly distributed between Level 3 and Level 4/5.

There were considerable differences between countries in performance at the different ISCED levels. At all education levels the proportion of Swedes who perform at the two lowest literacy levels on prose is much lower than in other countries. For example, in Sweden, those with second level, first stage education (ISCED 2) are much less likely to be at literacy Level 1 or 2 than are those with a similar level of education in other countries. Among those educated to upper secondary level (ISCED 3) the Netherlands distribution is very similar to that of Sweden which does not fit with the Netherlands performance at the other educational levels. This

probably reflects the application of ISCED in the Netherlands where courses which in other countries would be assigned to third level, non-university are included in upper secondary education.

(Figure 6.6, Table A6.2)

However, because of the strong association between education attainment level and literacy level, when education is held constant some of the differences between countries disappear. There was no significant difference between Britain and Canada, or either language group in Switzerland, in the proportion of those with lower secondary education (ISCED 2) who were at prose Level 1. Sweden, the Netherlands and Germany had significantly smaller proportions of those with this level of education at

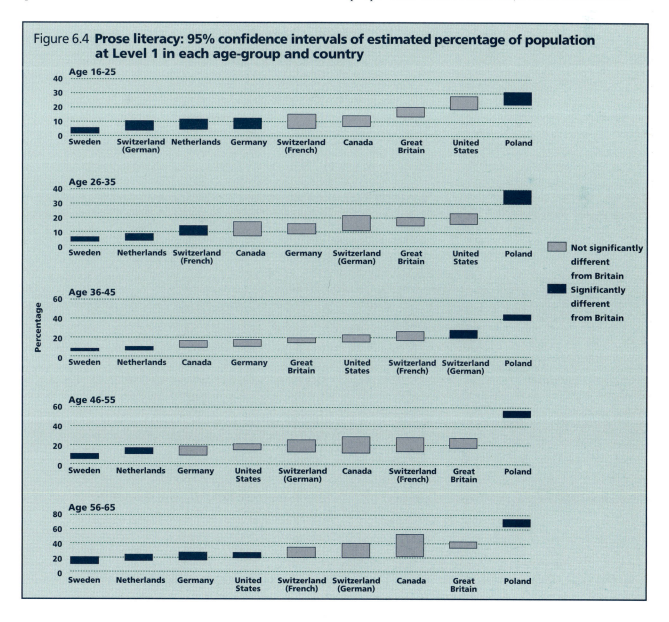

Figure 6.4 **Prose literacy: 95% confidence intervals of estimated percentage of population at Level 1 in each age-group and country**

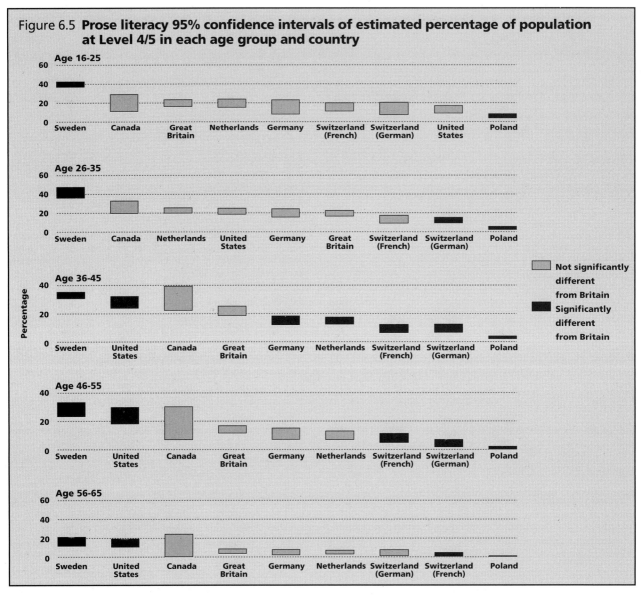

Figure 6.5 **Prose literacy 95% confidence intervals of estimated percentage of population at Level 4/5 in each age group and country**

the lowest level than Britain and the US had significantly higher proportions. Among those with education to second level, 2nd stage (ISCED 3) only Sweden, the Netherlands and Germany had significantly lower proportions at Level 1 than Britain. (The significant differences and confidence intervals for each education level and each prose level are shown in the annex Figures A6.1a - A6.1d).

At the higher literacy levels (Level 4/5) the variation in performance between Britain and other countries is even less clear. Except for those with primary education or lower, only Sweden had significantly higher proportions of people performing at Level 4/5 than Britain. Among those educated to second level, first stage (ISCED 2) a significantly higher proportion of Britons performed at Level 4/5 on

prose compared with those with the same level of educational attainment in the Netherlands, the US or Poland.

(Figure A6.1a-d, Table A6.2)

Having considered both age and education it is evident that the differing age and education distributions alone do not explain differences in the profile of literacy skills between countries. For example, although the Swedish population is very similar to that of Britain both in terms of the age profile of the labour force and the distribution of educational attainment, even across age-groups, the distribution of literacy skills in the two countries is quite different. Sweden has larger proportions of people performing at the higher literacy levels in each age-group and in each level of education compared

Table 6.5 Percentage of population at each document literacy level who are unemployed by country

	Level 1	Level 2	Level 3	Level 4/5	Level 1	Level 2	Level 3	Level 4/5
	Percentage who are unemployed				Standard errors*			
Canada	13	9	6	5	3.3	1.7	1.0	1.9
Germany	14	9	5	5	3.0	1.4	0.9	1.2
Great Britain	13	11	7	6	1.7	1.1	0.9	1.3
Netherlands	7	4	5	2	2.1	0.9	0.8	0.6
Poland	10	11	9	6	0.9	1.3	1.5	1.7
Sweden	12	7	5	5	2.1	0.7	0.6	0.5
Switzerland (French)	4	4	4	5	1.7	1.0	1.0	1.6
Switzerland (German)	3	2	2	3	1.6	1.0	0.9	1.1
United States	6	4	3	3	1.0	0.8	0.8	1.0

* The reported sample estimate can be said to be within 2 standard errors of the true population value with 95% confidence

with Britain. The opposite phenomenon is also evident when comparing Britain and the US. Both countries have similar age profiles and in all but the youngest age-group the US has higher proportions of the population who have third level education yet the literacy profiles of both countries are very similar. It is therefore difficult to know how to interpret these differences. When education is held constant there is no consistent pattern across countries other than Sweden who consistently outperforms those in other countries. Among those with lower secondary and upper secondary education, Germany has significantly lower proportions than Britain performing at Level 1. Some of these differences could be due to some of the anomalies associated with the ISCED (see Chapter 2, section 2.5) in that the application or interpretation of ISCED is not equivalent[8] although such anomalies are unlikely to account for all the observed differences.

(Figures 6.6, A6.1d, Table A6.2)

6.7 Literacy skills of the employed across different countries

In all countries apart from the French speaking part of Switzerland people performing at the lower literacy levels were more likely to be unemployed than those who performed at the higher literacy levels. In Britain, 13% of those at Level 1 and 11% of those at Level 2 on the document scale were unemployed compared with 7% and 6% of those who were at Level 3 or Level 4/5 respectively. These levels are similar to those in Canada, Sweden and Germany. The absolute proportion of those at each literacy level who are unemployed varies between countries,

reflecting the different unemployment rates. Although in most countries those at the lower literacy levels are more likely to be unemployed it is not necessarily true that raising the literacy levels of the unemployed would reduce unemployment. For example, at times when there is a large supply of jobs which do not require high literacy skills then having low skills will not necessarily be a barrier to employment. In times of job shortages however those with low skills may well find themselves unable to compete against more skilled workers.

(Table 6.5)

To look at it a slightly different way, Table 6.6 takes the employed and unemployed as the base and shows the percentage in each country at the different literacy levels using document literacy as the example. The percentage of the unemployed performing at the different literacy levels varied between countries and in most countries the unemployed were more likely to be at Level 1 on the document scale than the employed although the estimates for the unemployed are very imprecise for some countries. Canada, Germany, Britain and the US have large proportions of the unemployed population performing at Level 2 or lower on the document scale.

(Tables 6.6, A6.3)

6.8 Literacy skill and occupation

In Chapter 3 we saw that different occupational groups had quite different literacy levels in Britain and that some occupations were more homogenous than others in the skill profile of those holding those

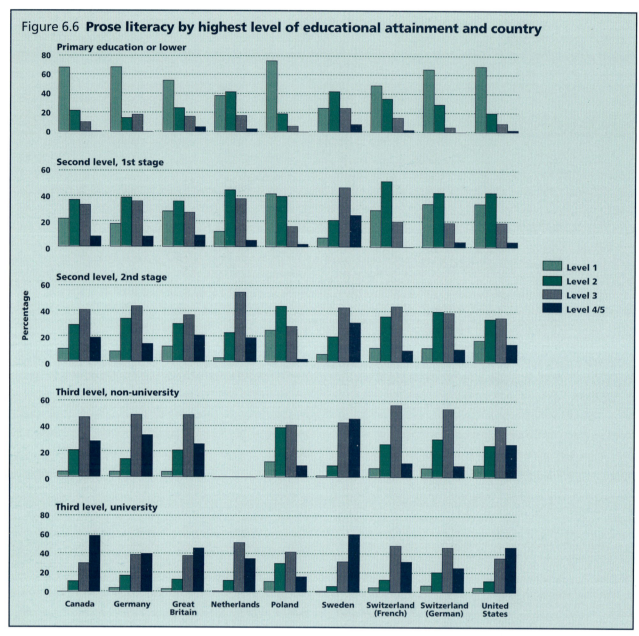

Figure 6.6 **Prose literacy by highest level of educational attainment and country**

jobs. It was also found that some occupational groups are more consistent than others across the different dimensions of literacy. As with other characteristics there are differences between countries in the distribution of occupation in the population. Table 6.7 shows the occupations of the employed in three countries. Germany, for example, has a much larger proportion of its workforce who are in craft and trade related occupations and a smaller proportion in managerial/professional occupations than either the UK or the Netherlands.

(Table 6.7)

Classifications of occupation take into account such factors as the minimum education level, qualifications and experience required to enter the different occupations and much of the variation in literacy skill by occupation is due to the underlying association with education. In all countries the expected relationship between literacy and occupation was observed with large proportions of managers/professionals and technicians performing at the higher literacy levels on all three scales. Among sales and service workers Canada, the US and to a lesser extent Britain had similar and more uniform distribution of literacy with similar proportions at each of the literacy levels while Germany had more centrally distributed skill levels and the Netherlands had a majority of such worker at Level 3. There were also considerable differences between countries in

Table 6.6   Document literacy level by employment status

	Level 1		Level 2		Level 3		Level 4/5		Total
	%	s.e.	%	s.e.	%	s.e.	%	s.e.	%
Employed persons									
Canada	12	1.0	24	2.5	35	1.4	30	2.2	100
Germany	5	0.9	31	1.5	42	1.1	22	1.2	100
Great Britain	16	0.9	26	1.2	34	1.2	23	1.4	100
Netherlands	6	0.8	22	0.8	48	1.1	24	1.1	100
Poland	41	1.9	31	1.5	21	1.2	7	0.6	100
Sweden	5	0.5	17	1.1	41	1.0	38	0.9	100
Switzerland (French)	12	1.3	30	2.2	41	1.8	17	1.4	100
Switzerland (German)	14	1.1	31	1.5	38	1.1	17	1.0	100
United States	18	0.7	26	1.2	34	1.1	23	1.3	100
Unemployed persons									
Canada	30	8.5	29	5.7	23	3.4	17	6.0	100
Germany	18	3.1	41	5.8	26	3.4	15	2.9	100
Great Britain	32	3.0	33	3.0	22	2.5	13	2.5	100
Netherlands	17	4.9	26	5.4	47	5.7	11	2.7	100
Poland	47	3.0	33	3.9	16	1.8	3	1.0	100
Sweden	12	2.1	23	1.6	35	2.6	30	2.4	100
Switzerland (French)	15	6.4	28	4.8	36	6.2	21	5.0	100
Switzerland (German)	24	10.7	23	10.0	35	12.3	17	7.5	100
United States	36	5.5	27	4.6	25	4.7	13	4.1	100

s.e. Standard error of the estimate. The reported sample estimate can be said to be within 2 standard errors of the true population value with 95% confidence

Table 6.7   Occupation (ISCO-88) by country
Persons in employment

Occupation	Legislators/ managers	Profess- ionals	Techni- cians	Clerks	Service and sales workers	Agriculture and fisheries workers	Craft and related trades workers	Plant and machine operators	Elementary occupations	Armed forces	Total
	%	%	%	%	%	%	%	%	%	%	%
Germany	6	11	19	13	11	2	19	8	11	1	100
Netherlands	12	16	18	12	13	2	11	8	7	1	100
UK	14	15	8	17	14	1	13	8	9	0	100

Source: Germany, Netherland, UK Labour Force Survey results 1994 EuroStat Luxembourg 1996 Table O45 p111

the distribution of literacy among skilled craft workers; the United States and Britain had similar distributions with large proportions of such workers at Levels 1 and 2. In both Germany and Sweden skilled craft workers were much more likely to perform at the higher literacy levels than were similar workers in other countries. The distribution of literacy among machine operators and assemblers and agriculture and primary occupations is quite different across countries. Both the Netherlands and Sweden have large proportions of such workers performing at Level 3 or higher whilst workers in these occupations in other countries were concentrated in the two lower literacy levels.

(Figure 6.7, Table A6.4)

6.9 Occupational demands for literacy skills

Respondents were asked how often they undertook various reading, writing and mathematics tasks as part of their job. The questions however did not try to ascertain the difficulty of the tasks they were required to perform and reading a letter or memo could conceivably represent a task from any literacy level. The proportion of the population who reported engaging in the various reading tasks was very similar in the three English speaking countries, Canada, the US and Britain except for manuals and reference books. In Canada, the US and Britain about 70% read letters or memos at least once week, 55-60% read reports, articles, magazines or journals and just under half read bills, invoices, spreadsheets or budget tables. Britons and Canadians were less likely to consult

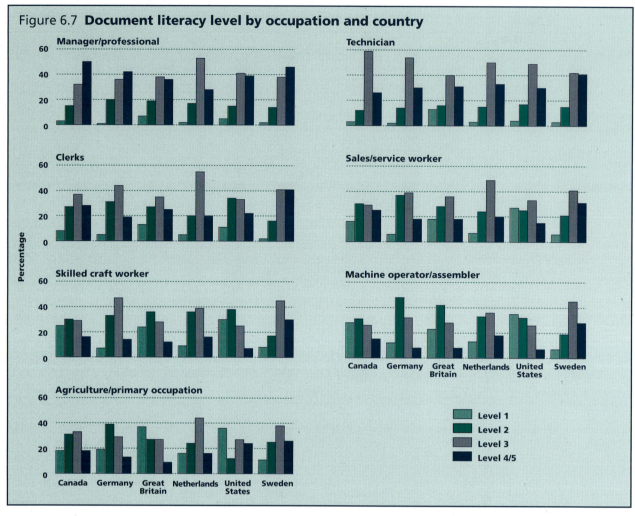

Figure 6.7 **Document literacy level by occupation and country**

manuals or reference books at least once a week as part of their job than were those in the US. The similarities between the age and education profiles of Britain and Sweden have already been mentioned in terms of how they fail to explain differences in the distribution of literacy skill between the two countries (section 6.5). On all reading activities a higher proportion of Swedes than Britons engaged in them at least once a week as part of their job. This may contribute to the better performance of the Swedes on the literacy measures.

(Table 6.8)

The extent to which similar jobs in different countries involve dealing with different types of literacy tasks varies. Figure 6.8 shows the percentage in each occupation who reported undertaking the various literacy reading activities at least once a week in Britain, Sweden and the US. The distributions for the US and Britain are very similar for most occupations on the reading tasks whereas in Sweden certain occupations show a higher frequency of use for some

tasks compared with the other countries. For example, in Sweden higher proportions of clerks and sales/service workers used manuals or reference books and reports, articles or journals at least once a week than did those in similar occupations in the other countries. Similarly, on the writing and mathematics tasks certain occupations are involved in these tasks on a more regular basis than others. One task that stands out in Figure 6.9 is the measurement or estimation of the size or weight of objects which is carried out most frequently by skilled craft workers in Britain, the US and Sweden.

(Figures 6.8 and 6.9, Tables A6.6 and A6.7)

Some occupations had very similar distributions across the three dimensions of literacy measured in the survey. Occupations that require high levels of education or that deal with a wide variety of literacy tasks on a regular basis tend to have this pattern, for example, managers/professionals, technicians and clerks. Occupations which have lower entry

Table 6.8 Percentage who reported engaging in each of several workplace reading activities at least once a week

People who had worked in the 12 months prior to interview

	Letters or memos	Reports, articles, magazines or journals	Manuals or reference books, including catalogues	Diagrams or schematics	Bills, invoices, spreadsheets or budget tables	Directions or instructions for medicines, recipes or other products
	Percentage engaging in each activity					
Switzerland (German)	81	70	56	32	61	17
Germany	80	67	61	51	62	32
Sweden	78	67	72	63	58	..
Switzerland (French)	73	72	50	38	57	19
Great Britain	73	59	54	37	47	30
United States	72	60	62	38	48	34
Canada	70	55	49	32	48	30
Netherlands	67	62	52	40	43	24
Poland	34	30	27	21	28	24

Figure 6.8 Percentage engaging in various reading tasks as part of their job at least once a week by occupation and country

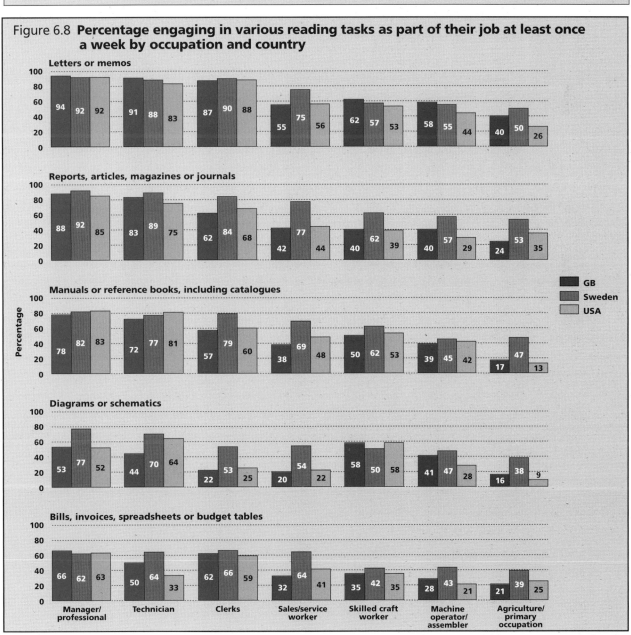

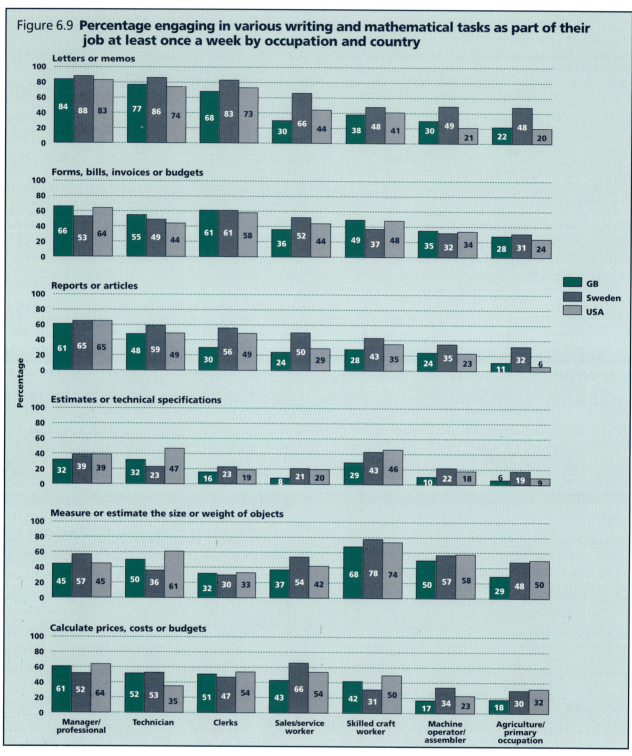

Figure 6.9 **Percentage engaging in various writing and mathematical tasks as part of their job at least once a week by occupation and country**

requirements or which do not involve such a wide variety of literacy skills such as skilled craft workers were in some countries less similar across the literacy dimensions. The training of skilled craft workers is quite different between countries, particularly in the entry requirements for apprenticeship and in the type and length of apprenticeship training provided. In Germany, the Netherlands and Sweden such workers performed better on document and quantitative

literacy than on prose whereas in Canada, the US and to a lesser extent in Britain there was more consistency in performance across the three scales. The profile of the literacy skills of machine operators and assemblers on the three scales was also quite different across countries. In most of the European countries these workers performed better on document and quantitative literacy relative to their performance on prose whereas in the two North

American countries there was more consistency in the performance of such workers on the three scales. In Canada however, these workers performed better on the prose scale than on the other two scales.

(Table A6.)

Notes and references

1 *Literacy, Economy and Society: Results of the first International Adult Literacy Survey.* OECD and Statistics Canada (1995)

2 To be published jointly by OECD and Statistics Canada.

3 See for example *Competitiveness Occasional Paper The Skills Audit : A Report from an Interdepartmental Group.* DfEE and Cabinet Office (1996) and Robinson Peter. *Literacy and Numeracy and Economic Performance.* Working Paper No. 888 Centre for Economic Performance London School of Economics (June 1997).

4 The standard error is one measure of the precision of a survey estimate simply as a result of using a sample rather than the whole population. The results obtained for any single sample may, by chance, vary from the true values for the population. The standard error allows calculation of the confidence intervals around the sample estimate which give an indication of the range in which the true population value is likely to fall. The interval is calculated as 1.96 times the standard error on either side of the estimated percentage or mean, since under a normal distribution, 95% of values lie within 1.96 standard errors of the mean value. If it were possible to repeat the survey under the same conditions many times, 95% of these confidence intervals would contain the population value.

5 In the British survey some questions were asked of more sub-groups than was required by the international survey design. In this chapter, the population on which the results are reported is the same as that used in the international report.

6 Keys W., Harris S. and Fernandes C. *Third International Mathematics and Science Study First National Report Part 1* NFER: (1996)

7 Under 25 year olds are excluded as a large proportion have not yet completed their education.

8 It is particularly difficult to establish exact equivalence of qualifications between different countries given that apart from differences in the structure of educational systems there are enormous differences between countries in the curricula, subject coverage and length of study to obtain what are thought to be equivalent qualifications. The ISCED is currently being reviewed.

Annex tables
A2.1-A2.11

Table A2.1 Literacy level by age and sex

	Level 1		Level 2		Level 3		Level 4/5		Total	Mean score	Base
	%	s.e.	%	s.e.	%	s.e.	%	s.e.	%		
Prose literacy											
Men											
16-25	18		31		34		17		100	270	*251*
26-35	18		27		36		19		100	274	*405*
36-45	16		29		34		21		100	276	*406*
46-55	20		27		35		18		100	269	*335*
56-65	37		37		20		6		100	238	*333*
Total	**21**		**30**		**32**		**17**		**100**	**267**	*1730*
Women											
16-25	16		30		32		22		100	279	*298*
26-35	18		31		32		19		100	274	*586*
36-45	19		28		32		21		100	274	*438*
46-55	25		31		34		10		100	260	*389*
56-65	40		36		18		6		100	236	*370*
Total	**22**		**31**		**30**		**16**		**100**	**266**	*2081*
All											
16-25	17	1.7	30	2.4	33	2.7	20	1.8	100	274	*549*
26-35	18	1.5	29	2.0	34	2.0	19	1.5	100	275	*991*
36-45	17	1.3	29	2.3	33	2.2	21	1.6	100	277	*844*
46-55	22	2.6	29	2.7	35	2.4	14	1.3	100	264	*724*
56-65	39	2.2	37	2.2	19	1.5	6	1.2	100	236	*703*
Total	**22**	**1.0**	**30**	**1.3**	**31**	**1.2**	**17**	**0.8**	**100**	**267**	*3811*
Document literacy											
Men											
16-25	15		25		35		25		100	278	*251*
26-35	18		21		33		28		100	284	*405*
36-45	16		23		32		29		100	284	*406*
46-55	20		24		32		25		100	277	*335*
56-65	34		36		23		8		100	239	*333*
Total	**20**		**25**		**31**		**24**		**100**	**274**	*1730*
Women											
16-25	21		28		33		18		100	273	*298*
26-35	20		30		31		19		100	270	*586*
36-45	22		25		32		20		100	269	*438*
46-55	29		32		31		9		100	253	*389*
56-65	47		29		19		5		100	227	*370*
Total	**27**		**29**		**30**		**15**		**100**	**260**	*2081*
All											
16-25	18	1.8	27	1.9	34	2.4	22	2.0	100	276	*549*
26-35	19	1.6	25	2.2	32	1.8	23	1.6	100	278	*991*
36-45	19	1.7	24	2.3	32	2.0	24	1.8	100	278	*844*
46-55	24	2.3	28	1.7	31	2.9	16	1.6	100	264	*724*
56-65	40	2.4	33	2.3	21	1.3	6	1.0	100	233	*703*
Total	**23**	**1.0**	**27**	**1.0**	**31**	**1.0**	**19**	**1.0**	**100**	**268**	*3811*
Quantitative literacy											
Men											
16-25	17		31		32		20		100	271	*251*
26-35	15		26		27		32		100	287	*405*
36-45	14		22		33		32		100	291	*406*
46-55	17		21		35		27		100	284	*335*
56-65	29		35		25		11		100	251	*333*
Total	**18**		**27**		**30**		**25**		**100**	**278**	*1730*
Women											
16-25	28		27		33		12		100	260	*298*
26-35	24		30		33		14		100	264	*586*
36-45	24		27		32		17		100	266	*438*
46-55	31		31		30		8		100	250	*389*
56-65	42		32		22		4		100	232	*370*
Total	**29**		**29**		**30**		**12**		**100**	**256**	*2081*
All											
16-25	22	1.7	29	2.5	33	2.4	16	2.1	100	265	*549*
26-35	20	1.7	28	2.0	30	2.0	23	1.6	100	277	*991*
36-45	19	1.2	24	2.0	32	1.9	25	1.5	100	279	*844*
46-55	24	2.3	26	2.0	32	2.8	17	1.7	100	266	*724*
56-65	35	2.3	34	2.5	23	1.8	8	0.9	100	240	*703*
Total	**23**	**0.9**	**28**	**1.0**	**30**	**0.9**	**19**	**1.0**	**100**	**267**	*3811*

s.e. = Standard error of the estimate. The reported sample estimate can be said to be within 2 standard errors of the true population value with 95% confidence. The standard errors were not available for men and women separately.

Table A2.1A **Literacy level by age and sex (cumulative percentages)**

	Level 4/5	Level 3 or above	Level 2 or above	Level 1 or above
		Cumulative %		
Prose literacy				
Men				
16-25	17	51	82	100
26-35	19	55	82	100
36-45	21	55	84	100
46-55	18	53	80	100
56-65	6	26	63	100
Total	17	49	79	100
Women				
16-25	22	54	84	100
26-35	19	51	82	100
36-45	21	53	81	100
46-55	10	44	75	100
56-65	6	24	60	100
Total	16	46	77	100
All				
16-25	20	53	83	100
26-35	19	53	82	100
36-45	21	54	83	100
46-55	14	49	78	100
56-65	6	25	62	100
Total	17	48	78	100
Document literacy				
Men				
16-25	25	60	85	100
26-35	28	61	82	100
36-45	29	61	84	100
46-55	25	57	81	100
56-65	8	31	67	100
Total	24	55	80	100
Women				
16-25	18	51	79	100
26-35	19	50	80	100
36-45	20	52	77	100
46-55	9	40	72	100
56-65	5	24	53	100
Total	15	45	74	100
All				
16-25	22	56	83	100
26-35	23	55	80	100
36-45	24	56	80	100
46-55	16	47	75	100
56-65	6	27	60	100
Total	19	50	77	100
Quantitative literacy				
Men				
16-25	20	52	83	100
26-35	32	59	85	100
36-45	32	65	87	100
46-55	27	62	83	100
56-65	11	36	71	100
Total	25	55	82	100
Women				
16-25	12	45	72	100
26-35	14	47	77	100
36-45	17	49	76	100
46-55	8	38	69	100
56-65	4	26	58	100
Total	12	42	71	100
All				
16-25	16	49	78	100
26-35	23	53	81	100
36-45	25	57	81	100
46-55	17	49	75	100
56-65	8	31	65	100
Total	19	49	77	100

Table A2.2 **Literacy level by level of highest qualification and sex**

	Level 1	Level 2	Level 3	Level 4/5	Total	Mean score	Base
	%	%	%	%	%		
Prose literacy							
Men							
A-Levels, vocational level 3 & above	4	22	43	31	100	303	*734*
Higher Education & professional/vocational equivalents	2	17	42	39	100	311	*493*
Degree or equivalent	2	14	39	45	100	315	*314*
Other Higher Education below degree level	2	24	48	27	100	303	*179*
A-Levels, vocational level 3 & equivalents	6	28	45	21	100	292	*241*
Other qualifications	19	37	33	12	100	264	*629*
Trade apprenticeships	24	41	28	7	100	256	*248*
GCSE/O Level grade A*-C, vocational level 2 & equivalents	7	27	45	21	100	288	*230*
Qualifications below level 2	21	49	24	6	100	253	*89*
Other qualifications - level unknown	42	36	18	4	100	223	*62*
No qualifications	52	30	15	3	100	215	*367*
Total	**21**	**30**	**32**	**17**	**100**	**267**	***1730***
Women							
A-Levels, vocational level 3 & above	5	17	43	36	100	307	*674*
Higher Education & professional/vocational equivalents	5	15	44	36	100	309	*461*
Degree or equivalent	4	11	38	47	100	316	*225*
Other Higher Education below degree level	6	19	49	26	100	301	*236*
A-Levels, vocational level 3 & equivalents	5	20	41	34	100	303	*213*
Other qualifications	18	37	32	13	100	269	*784*
Trade apprenticeships	31	42	23	3	100	246	*87*
GCSE/O Level grade A*-C, vocational level 2 & equivalents	10	34	37	19	100	283	*438*
Qualifications below level 2	17	48	29	6	100	264	*186*
Other qualifications - level unknown	52	27	18	3	100	224	*73*
No qualifications	45	35	17	3	100	225	*623*
Total	**22**	**31**	**30**	**16**	**100**	**266**	***2081***
All							
A-Levels, vocational level 3 & above	4	20	43	33	100	305	*1408*
Higher Education & professional/vocational equivalents	4	16	43	38	100	310	*954*
Degree or equivalent	3	13	39	46	100	316	*539*
Other Higher Education below degree level	4	21	49	26	100	302	*415*
A-Levels, vocational level 3 & equivalents	5	25	43	26	100	296	*454*
Other qualifications	18	37	32	13	100	267	*1413*
Trade apprenticeships	26	42	27	6	100	253	*335*
GCSE/O Level grade A*-C, vocational level 2 & equivalents	9	31	40	20	100	285	*668*
Qualifications below level 2	19	48	27	6	100	260	*275*
Other qualifications - level unknown	47	32	18	3	100	223	*135*
No qualifications	48	33	16	3	100	221	*990*
Total	**22**	**30**	**31**	**17**	**100**	**267**	***3811***

Table A2.2 (continued) **Literacy level by level of highest qualification and sex**

	Level 1	Level 2	Level 3	Level 4/5	Total	Mean score	Base
	%	%	%	%	%		
Document literacy							
Men							
A-Levels, vocational level 3 & above	5	16	37	42	100	313	734
Higher Education & professional/vocational equivalents	3	12	36	49	100	322	493
Degree or equivalent	3	9	33	55	100	326	314
Other Higher Education below degree level	3	19	43	35	100	313	179
A-Levels, vocational level 3 & equivalents	7	20	39	34	100	300	241
Other qualifications	18	30	34	17	100	273	629
Trade apprenticeships	25	34	31	9	100	260	248
GCSE/O Level grade A*-C, vocational level 2 & equivalents	8	22	40	31	100	299	230
Qualifications below level 2	12	43	31	13	100	268	89
Other qualifications - level unknown	44	30	25	1	100	224	62
No qualifications	46	32	17	5	100	218	367
Total	**20**	**25**	**31**	**24**	**100**	**274**	**1730**
Women							
A-Levels, vocational level 3 & above	8	19	43	31	100	300	674
Higher Education & professional/vocational equivalents	8	19	41	32	100	301	461
Degree or equivalent	5	18	37	41	100	310	225
Other Higher Education below degree level	11	21	45	22	100	292	236
A-Levels, vocational level 3 & equivalents	7	18	45	29	100	299	213
Other qualifications	22	33	33	12	100	265	784
Trade apprenticeships	44	29	24	3	100	239	87
GCSE/O Level grade A*-C, vocational level 2 & equivalents	12	32	38	18	100	280	438
Qualifications below level 2	22	44	29	5	100	257	186
Other qualifications - level unknown	55	16	24	4	100	217	73
No qualifications	51	32	13	4	100	218	623
Total	**27**	**29**	**30**	**15**	**100**	**260**	**2081**
All							
A-Levels, vocational level 3 & above	6	17	40	37	100	307	1408
Higher Education & professional/vocational equivalents	5	16	39	41	100	312	954
Degree or equivalent	4	12	35	50	100	320	539
Other Higher Education below degree level	7	20	45	28	100	301	415
A-Levels, vocational level 3 & equivalents	7	20	41	32	100	300	454
Other qualifications	20	32	34	15	100	269	1413
Trade apprenticeships	30	33	30	8	100	255	335
GCSE/O Level grade A*-C, vocational level 2 & equivalents	10	28	39	23	100	288	668
Qualifications below level 2	18	43	30	9	100	262	275
Other qualifications - level unknown	49	23	25	3	100	221	135
No qualifications	49	32	15	4	100	218	990
Total	**23**	**27**	**31**	**19**	**100**	**268**	**3811**

Table A2.2 (continued) **Literacy level by level of highest qualification and sex**

	Level 1	Level 2	Level 3	Level 4/5	Total	Mean score	Base
	%	%	%	%	%		
Quantitative literacy							
Men							
A-Levels, vocational level 3 & above	4	18	32	46	100	316	734
Higher Education & professional/vocational equivalents	2	12	28	57	100	328	493
Degree or equivalent	2	9	24	65	100	335	314
Other Higher Education below degree level	3	18	37	42	100	315	179
A-Levels, vocational level 3 & equivalents	6	24	38	32	100	301	241
Other qualifications	16	31	35	18	100	275	629
Trade apprenticeships	20	36	35	9	100	264	248
GCSE/O Level grade A*-C, vocational level 2 & equivalents	6	25	37	32	100	300	230
Qualifications below level 2	20	33	34	13	100	269	89
Other qualifications - level unknown	37	32	28	2	100	231	62
No qualifications	42	33	20	5	100	224	367
Total	**18**	**27**	**30**	**25**	**100**	**278**	**1730**
Women							
A-Levels, vocational level 3 & above	8	20	43	29	100	297	674
Higher Education & professional/vocational equivalents	7	21	42	30	100	299	461
Degree or equivalent	5	16	38	41	100	311	225
Other Higher Education below degree level	9	25	46	19	100	287	236
A-Levels, vocational level 3 & equivalents	9	20	45	26	100	294	213
Other qualifications	27	31	34	8	100	258	784
Trade apprenticeships	44	31	21	3	100	237	87
GCSE/O Level grade A*-C, vocational level 2 & equivalents	18	28	43	11	100	272	438
Qualifications below level 2	28	42	26	4	100	252	186
Other qualifications - level unknown	55	21	21	3	100	218	73
No qualifications	51	35	13	2	100	215	623
Total	**29**	**29**	**30**	**12**	**100**	**256**	**2081**
All							
A-Levels, vocational level 3 & above	6	19	37	39	100	308	1408
Higher Education & professional/vocational equivalents	4	16	34	45	100	315	954
Degree or equivalent	3	12	29	56	100	325	539
Other Higher Education below degree level	7	22	42	30	100	300	415
A-Levels, vocational level 3 & equivalents	7	23	41	30	100	298	454
Other qualifications	21	31	35	13	100	267	1413
Trade apprenticeships	26	35	31	8	100	257	335
GCSE/O Level grade A*-C, vocational level 2 & equivalents	13	27	41	19	100	283	668
Qualifications below level 2	25	38	29	8	100	259	275
Other qualifications - level unknown	46	27	25	3	100	224	135
No qualifications	47	34	16	3	100	219	990
Total	**23**	**28**	**30**	**19**	**100**	**267**	**3811**

Table A2.2a **Literacy level by level of highest qualification and sex (cumulative percentages)**

	Level 4/5	Level 3 or above	Level 2 or above	Level 1 or above
		Cumulative %		
Prose literacy				
Men				
A-Levels, vocational level 3 & above	31	74	96	100
Higher Education & professional/vocational equivalents	39	81	98	100
Degree or equivalent	45	84	98	100
Other Higher Education below degree level	27	75	99	100
A-Levels, vocational level 3 & equivalents	21	66	94	100
Other qualifications	12	45	82	100
Trade apprenticeships	7	35	76	100
GCSE/O Level grade A*-C, vocational level 2 & equivalents	21	66	93	100
Qualifications below level 2	6	30	79	100
Other qualifications - level unknown	4	22	58	100
No qualifications	3	18	48	100
Total	**17**	**49**	**79**	**100**
Women				
A-Levels, vocational level 3 & above	36	79	96	100
Higher Education & professional/vocational equivalents	36	80	95	100
Degree or equivalent	47	85	96	100
Other Higher Education below degree level	26	75	94	100
A-Levels, vocational level 3 & equivalents	34	75	95	100
Other qualifications	13	45	82	100
Trade apprenticeships	3	26	68	100
GCSE/O Level grade A*-C, vocational level 2 & equivalents	19	56	90	100
Qualifications below level 2	6	35	83	100
Other qualifications - level unknown	3	21	48	100
No qualifications	3	20	55	100
Total	**16**	**46**	**77**	**100**
All				
A-Levels, vocational level 3 & above	33	76	96	100
Higher Education & professional/vocational equivalents	38	81	97	100
Degree or equivalent	46	85	98	100
Other Higher Education below degree level	26	75	96	100
A-Levels, vocational level 3 & equivalents	26	69	94	100
Other qualifications	13	45	82	100
Trade apprenticeships	6	33	75	100
GCSE/O Level grade A*-C, vocational level 2 & equivalents	20	60	91	100
Qualifications below level 2	6	33	81	100
Other qualifications - level unknown	3	21	53	100
No qualifications	3	19	52	100
Total	**17**	**48**	**78**	**100**

Table A2.2a (continued) **Literacy level by level of highest qualification and sex (cumulative percentages)**

	Level 4/5	Level 3 or above	Level 2 or above	Level 1 or above
		Cumulative %		
Document literacy				
Men				
A-Levels, vocational level 3 & above	42	79	95	100
Higher Education & professional/vocational equivalents	49	85	97	100
Degree or equivalent	55	88	97	100
Other Higher Education below degree level	35	78	97	100
A-Levels, vocational level 3 & equivalents	34	73	93	100
Other qualifications	17	51	81	100
Trade apprenticeships	9	40	74	100
GCSE/O Level grade A*-C, vocational level 2 & equivalents	31	71	93	100
Qualifications below level 2	13	44	87	100
Other qualifications - level unknown	1	26	56	100
No qualifications	5	22	54	100
Total	**24**	**55**	**80**	**100**
Women				
A-Levels, vocational level 3 & above	31	74	93	100
Higher Education & professional/vocational equivalents	32	73	92	100
Degree or equivalent	41	78	96	100
Other Higher Education below degree level	22	67	88	100
A-Levels, vocational level 3 & equivalents	29	74	92	100
Other qualifications	12	45	78	100
Trade apprenticeships	3	27	56	100
GCSE/O Level grade A*-C, vocational level 2 & equivalents	18	56	88	100
Qualifications below level 2	5	34	78	100
Other qualifications - level unknown	4	28	44	100
No qualifications	4	17	49	100
Total	**15**	**45**	**74**	**100**
All				
A-Levels, vocational level 3 & above	37	77	94	100
Higher Education & professional/vocational equivalents	41	80	96	100
Degree or equivalent	50	85	97	100
Other Higher Education below degree level	28	73	93	100
A-Levels, vocational level 3 & equivalents	32	73	93	100
Other qualifications	15	49	81	100
Trade apprenticeships	8	38	71	100
GCSE/O Level grade A*-C, vocational level 2 & equivalents	23	62	90	100
Qualifications below level 2	9	39	82	100
Other qualifications - level unknown	3	28	51	100
No qualifications	4	19	51	100
Total	**19**	**50**	**77**	**100**

Table A2.2a (continued) **Literacy level by level of highest qualification and sex (cumulative percentages)**

	Level 4/5	Level 3 or above	Level 2 or above	Level 1 or above
		Cumulative %		
Quantitative literacy				
Men				
A-Levels, vocational level 3 & above	46	78	96	100
Higher Education & professional/vocational equivalents	57	85	97	100
Degree or equivalent	65	89	98	100
Other Higher Education below degree level	42	79	97	100
A-Levels, vocational level 3 & equivalents	32	70	94	100
Other qualifications	18	53	84	100
Trade apprenticeships	9	44	80	100
GCSE/O Level grade A*-C, vocational level 2 & equivalents	32	69	94	100
Qualifications below level 2	13	47	80	100
Other qualifications - level unknown	2	30	62	100
No qualifications	5	25	58	100
Total	**25**	**55**	**82**	**100**
Women				
A-Levels, vocational level 3 & above	29	72	92	100
Higher Education & professional/vocational equivalents	30	72	93	100
Degree or equivalent	41	79	95	100
Other Higher Education below degree level	19	65	90	100
A-Levels, vocational level 3 & equivalents	26	71	91	100
Other qualifications	8	42	73	100
Trade apprenticeships	3	24	55	100
GCSE/O Level grade A*-C, vocational level 2 & equivalents	11	54	82	100
Qualifications below level 2	4	30	72	100
Other qualifications - level unknown	3	24	45	100
No qualifications	2	15	50	100
Total	**12**	**42**	**71**	**100**
All				
A-Levels, vocational level 3 & above	39	76	95	100
Higher Education & professional/vocational equivalents	45	79	95	100
Degree or equivalent	56	85	97	100
Other Higher Education below degree level	30	72	94	100
A-Levels, vocational level 3 & equivalents	30	71	94	100
Other qualifications	13	48	79	100
Trade apprenticeships	8	39	74	100
GCSE/O Level grade A*-C, vocational level 2 & equivalents	19	60	87	100
Qualifications below level 2	8	37	75	100
Other qualifications - level unknown	3	28	55	100
No qualifications	3	19	53	100
Total	**19**	**49**	**77**	**100**

Table A2.3 **Literacy level by highest level of educational attainment and sex**

	Level 1		Level 2		Level 3		Level 4/5		Total	Mean score	Base
	%	s.e.	%	s.e.	%	s.e.	%	s.e.	%		
Prose literacy											
Men											
Primary education or lower	54	6.7	31	4.8	12	4.1	2	2.4	100	204	89
Second level, 1st stage	30	2.0	34	1.7	27	2.3	9	1.2	100	250	726
Second level, 2nd stage	12	2.0	32	2.5	40	2.8	17	2.4	100	278	415
Third level, non-university	2	0.9	26	3.5	47	4.4	26	3.7	100	301	187
Third level, university	2	1.1	14	2.6	39	3.4	45	3.7	100	316	313
Total	**21**	**1.0**	**30**	**1.2**	**32**	**1.3**	**17**	**1.1**	**100**	**267**	**1730**
Women											
Primary education or lower	53	6.4	19	5.7	20	6.2	7	3.6	100	208	113
Second level, 1st stage	27	2.0	38	2.3	27	1.9	9	0.8	100	256	1196
Second level, 2nd stage	13	2.4	27	3.2	31	3.9	29	5.1	100	287	313
Third level, non-university	6	2.3	18	3.5	50	3.8	26	3.7	100	300	234
Third level, university	4	1.8	11	2.0	38	3.8	47	3.6	100	318	225
Total	**22**	**1.7**	**31**	**1.8**	**30**	**1.6**	**16**	**1.0**	**100**	**267**	**2081**
All											
Primary education or lower	54	4.2	25	4.0	16	3.1	5	2.3	100	206	202
Second level, 1st stage	28	1.5	36	1.6	27	1.6	9	0.7	100	253	1922
Second level, 2nd stage	12	1.6	30	2.1	37	2.0	21	2.0	100	281	728
Third level, non-university	4	1.3	22	2.7	49	2.8	26	2.6	100	300	421
Third level, university	3	0.9	13	1.7	38	2.4	46	2.3	100	317	538
Total	**22**	**1.0**	**30**	**1.3**	**31**	**1.2**	**17**	**0.8**	**100**	**267**	**3811**
Document literacy											
Men											
Primary education or lower	50	6.5	26	5.1	19	5.1	5	2.5	100	209	89
Second level, 1st stage	28	1.8	30	1.3	29	2.0	14	1.6	100	256	726
Second level, 2nd stage	11	1.7	28	2.5	35	3.0	26	3.0	100	287	415
Third level, non-university	3	1.2	20	4.0	43	5.4	34	4.6	100	309	187
Third level, university	3	1.1	9	2.0	33	4.1	55	4.3	100	325	313
Total	**20**	**1.0**	**25**	**1.1**	**31**	**1.2**	**24**	**1.5**	**100**	**274**	**1730**
Women											
Primary education or lower	52	7.5	24	6.5	15	5.2	9	3.7	100	207	113
Second level, 1st stage	32	2.0	33	1.5	27	1.5	8	0.9	100	250	1196
Second level, 2nd stage	16	2.4	23	2.9	36	4.7	25	4.5	100	282	313
Third level, non-university	11	2.9	21	3.1	47	4.2	21	4.0	100	290	234
Third level, university	5	1.9	18	2.6	37	3.4	41	3.2	100	313	225
Total	**27**	**1.6**	**29**	**1.4**	**30**	**1.6**	**15**	**1.1**	**100**	**261**	**2081**
All											
Primary education or lower	51	4.6	25	3.9	17	3.4	7	2.3	100	208	202
Second level, 1st stage	30	1.5	32	1.1	27	1.4	11	1.0	100	252	1922
Second level, 2nd stage	13	1.5	26	1.9	35	2.5	26	2.7	100	285	728
Third level, non-university	7	1.6	21	2.3	45	3.2	27	2.7	100	299	421
Third level, university	3	1.0	12	1.4	35	2.7	50	2.7	100	320	538
Total	**23**	**1.0**	**27**	**1.0**	**31**	**1.0**	**19**	**1.0**	**100**	**268**	**3811**
Quantitative literacy											
Men											
Primary education or lower	49	6.1	31	5.3	17	5.3	2	2.4	100	214	89
Second level, 1st stage	25	2.2	31	2.1	30	1.8	15	1.5	100	260	726
Second level, 2nd stage	10	1.7	29	2.6	36	2.2	25	2.6	100	289	415
Third level, non-university	3	1.2	20	3.5	37	5.0	41	4.0	100	312	187
Third level, university	2	0.8	9	2.0	23	3.7	65	4.0	100	333	313
Total	**18**	**1.2**	**27**	**1.3**	**30**	**1.2**	**25**	**1.4**	**100**	**278**	**1730**
Women											
Primary education or lower	51	5.5	35	5.7	10	3.6	5	2.9	100	202	113
Second level, 1st stage	35	1.6	32	1.7	28	1.9	5	0.8	100	244	1196
Second level, 2nd stage	19	3.1	23	3.5	36	4.2	22	3.6	100	277	313
Third level, non-university	11	3.0	25	3.3	45	3.9	20	3.9	100	286	234
Third level, university	5	1.7	16	2.7	38	4.0	41	3.3	100	313	225
Total	**29**	**1.5**	**29**	**1.4**	**30**	**1.5**	**12**	**1.2**	**100**	**256**	**2081**
All											
Primary education or lower	50	3.8	33	3.9	14	2.8	3	2.0	100	208	202
Second level, 1st stage	30	1.2	32	1.5	29	1.4	9	0.8	100	251	1922
Second level, 2nd stage	13	1.6	27	2.3	36	2.0	24	2.4	100	285	728
Third level, non-university	7	1.6	22	2.4	41	2.7	29	2.9	100	298	421
Third level, university	3	0.8	12	1.4	29	2.6	56	2.7	100	326	538
Total	**23**	**0.9**	**28**	**1.0**	**30**	**0.9**	**19**	**1.0**	**100**	**267**	**3811**

s.e. = Standard error of the estimate. The reported sample estimate can be said to be within 2 standard errors of the true population value with 95% confidence.

Table A2. 4 **Literacy level by economic activity status and sex**

	Level 1	Level 2	Level 3	Level 4/5	Total	Mean score	Base
	%	%	%	%	%		
Prose literacy							
Men							
Employed	15	29	36	20	100	278	1295
Unemployed	40	25	28	7	100	236	175
Student	10	32	36	22	100	*	36
Home duties	[9]	[7]	[1]	[0]	[17]	*	17
Retired	27	29	30	14	100	254	84
Other inactive	54	37	8	2	100	195	123
Total	**21**	**30**	**32**	**17**	**100**	**267**	**1730**
Women							
Employed	16	31	34	19	100	278	1320
Unemployed	24	30	29	17	100	267	183
Student	3	24	38	35	100	*	38
Home duties	35	33	25	7	100	240	277
Retired	44	36	17	3	100	231	134
Other inactive	52	30	13	5	100	211	129
Total	**22**	**31**	**30**	**16**	**100**	**266**	**2081**
All							
Employed	16	30	35	20	100	278	2615
Unemployed	32	28	28	12	100	251	358
Student	7	28	37	28	100	290	74
Home duties	36	35	23	7	100	239	294
Retired	36	33	23	8	100	241	218
Other inactive	53	34	10	3	100	203	252
Total	**22**	**30**	**31**	**17**	**100**	**267**	**3811**
Document literacy							
Men							
Employed	14	25	34	28	100	287	1295
Unemployed	37	27	27	10	100	238	175
Student	10	27	31	32	100	*	36
Home duties	[8]	[6]	[3]	[0]	[17]	*	17
Retired	31	22	29	18	100	254	84
Other inactive	57	30	11	2	100	193	123
Total	**20**	**25**	**31**	**24**	**100**	**274**	**1730**
Women							
Employed	20	28	34	18	100	272	1320
Unemployed	27	39	18	16	100	261	183
Student	3	22	51	24	100	*	38
Home duties	41	29	24	6	100	235	277
Retired	52	29	17	2	100	220	134
Other inactive	60	25	12	4	100	204	129
Total	**27**	**29**	**30**	**15**	**100**	**260**	**2081**
All							
Employed	17	26	34	23	100	281	2615
Unemployed	32	33	22	13	100	249	358
Student	7	24	41	28	100	294	74
Home duties	41	30	23	6	100	234	294
Retired	43	26	22	9	100	234	218
Other inactive	58	28	11	3	100	198	252
Total	**23**	**27**	**31**	**19**	**100**	**268**	**3811**
Quantitative literacy							
Men							
Employed	12	26	32	30	100	291	1295
Unemployed	37	26	27	10	100	241	175
Student	16	30	30	25	100	*	36
Home duties	[6]	[6]	[5]	[0]	[17]	*	17
Retired	25	24	30	21	100	264	84
Other inactive	53	26	18	3	100	204	123
Total	**18**	**27**	**30**	**25**	**100**	**278**	**1730**
Women							
Employed	22	29	35	14	100	268	1320
Unemployed	29	36	27	7	100	252	183
Student	6	36	35	24	100	*	38
Home duties	44	28	22	5	100	228	277
Retired	50	30	18	2	100	222	134
Other inactive	58	25	14	4	100	207	129
Total	**29**	**29**	**30**	**12**	**100**	**256**	**2081**
All							
Employed	17	27	33	23	100	280	2615
Unemployed	33	31	27	9	100	246	358
Student	11	32	32	24	100	282	74
Home duties	43	29	23	5	100	229	294
Retired	40	27	23	10	100	240	218
Other inactive	55	25	16	4	100	206	252
Total	**23**	**28**	**30**	**19**	**100**	**267**	**3811**

Table A2. 4a **Literacy level by economic activity status and sex (cumulative percentages)**

	Level 4/5	Level 3 or above	Level 2 or above	Level 1 or above
		Cumulative %		
Prose literacy				
Men				
Employed	20	56	85	100
Unemployed	7	35	60	100
Student	22	58	90	100
Home duties	..	..	..	..
Retired	14	44	73	100
Other inactive	2	10	47	100
Total	**17**	**49**	**79**	**100**
Women				
Employed	19	53	84	100
Unemployed	17	46	76	100
Student	35	73	97	100
Home duties	7	32	65	100
Retired	3	20	56	100
Other inactive	5	18	48	100
Total	**16**	**46**	**77**	**100**
All				
Employed	20	55	85	100
Unemployed	12	40	68	100
Student	28	65	93	100
Home duties	7	30	65	100
Retired	8	31	64	100
Other inactive	3	13	47	100
Total	**17**	**48**	**78**	**100**
Document literacy				
Men				
Employed	28	62	87	100
Unemployed	10	37	64	100
Student	32	63	90	100
Home duties	..	..	..	..
Retired	18	47	69	100
Other inactive	2	13	43	100
Total	**24**	**55**	**80**	**100**
Women				
Employed	18	52	80	100
Unemployed	16	34	73	100
Student	24	75	97	100
Home duties	6	30	59	100
Retired	2	19	48	100
Other inactive	4	16	41	100
Total	**15**	**45**	**74**	**100**
All				
Employed	23	57	83	100
Unemployed	13	35	68	100
Student	28	69	93	100
Home duties	6	29	59	100
Retired	9	31	57	100
Other inactive	3	14	42	100
Total	**19**	**50**	**77**	**100**
Quantitative literacy				
Men				
Employed	30	62	88	100
Unemployed	10	37	63	100
Student	25	55	85	100
Home duties	..	..	..	..
Retired	21	51	75	100
Other inactive	3	21	47	100
Total	**25**	**55**	**82**	**100**
Women				
Employed	14	49	78	100
Unemployed	7	34	70	100
Student	24	59	95	100
Home duties	5	27	55	100
Retired	2	20	50	100
Other inactive	4	18	43	100
Total	**12**	**42**	**71**	**100**
All				
Employed	23	56	83	100
Unemployed	9	36	67	100
Student	24	56	88	100
Home duties	5	28	57	100
Retired	10	33	60	100
Other inactive	4	20	45	100
Total	**19**	**49**	**77**	**100**

Table A2. 5 **Literacy level by economic activity status and highest educational qualification**
All except full-time students

	Level 1	Level 2	Level 3	Level 4/5	Total	Mean score	Base
	%	%	%	%	%		
Prose literacy							
No qualifications							
Employed	41	35	20	4	100	235	487
Unemployed	57	28	11	5	100	215	109
Inactive	58	31	10	1	100	198	378
Total	**49**	**33**	**15**	**3**	**100**	**220**	**974**
Other qualifications							
Employed	15	37	34	14	100	273	994
Unemployed	26	30	34	10	100	257	155
Inactive	31	40	24	5	100	244	243
Total	**18**	**37**	**33**	**12**	**100**	**267**	**1392**
A levels, vocational level 3 and above							
Employed	4	19	44	34	100	306	1134
Unemployed	7	21	45	27	100	295	94
Inactive	11	28	35	26	100	289	143
Total	**4**	**20**	**43**	**33**	**100**	**304**	**1371**
All							
Employed	16	30	35	20	100	278	2615
Unemployed	32	28	28	12	100	251	358
Inactive	42	34	18	6	100	227	764
Total	**22**	**30**	**31**	**17**	**100**	**267**	**3737**
Document literacy							
No qualifications							
Employed	40	36	19	6	100	234	487
Unemployed	57	29	8	5	100	212	109
Inactive	63	26	10	1	100	191	378
Total	**50**	**31**	**14**	**4**	**100**	**217**	**974**
Other qualifications							
Employed	16	31	37	17	100	276	994
Unemployed	23	41	26	10	100	255	155
Inactive	38	32	25	5	100	239	243
Total	**20**	**32**	**34**	**14**	**100**	**268**	**1392**
A levels, vocational level 3 and above							
Employed	5	16	40	39	100	310	1134
Unemployed	11	21	37	31	100	293	94
Inactive	15	28	35	23	100	285	143
Total	**6**	**17**	**39**	**37**	**100**	**307**	**1371**
All							
Employed	17	26	34	23	100	281	2615
Unemployed	32	33	22	13	100	249	358
Inactive	47	28	19	6	100	221	764
Total	**23**	**27**	**31**	**19**	**100**	**267**	**3737**
Quantitative literacy							
No qualifications							
Employed	38	38	19	4	100	236	487
Unemployed	56	32	11	2	100	208	109
Inactive	60	26	13	0	100	195	378
Total	**48**	**33**	**16**	**3**	**100**	**218**	**974**
Other qualifications							
Employed	17	30	37	16	100	274	994
Unemployed	26	38	32	4	100	251	155
Inactive	39	32	24	4	100	239	243
Total	**22**	**31**	**34**	**13**	**100**	**266**	**1392**
A levels, vocational level 3 and above							
Employed	4	18	36	41	100	311	1134
Unemployed	13	16	43	29	100	295	94
Inactive	15	20	38	28	100	290	143
Total	**6**	**18**	**37**	**39**	**100**	**308**	**1371**
All							
Employed	17	27	33	23	100	280	2615
Unemployed	33	31	27	9	100	246	358
Inactive	46	27	21	6	100	224	764
Total	**23**	**28**	**30**	**19**	**100**	**267**	**3737**

Table A2. 6 **Literacy level by social class**

Social class	Level 1	Level 2	Level 3	Level 4/5	Total	Mean score	Base
	%	%	%	%	%		
Prose literacy							
I and II	9	21	41	28	100	294	1254
III (non-manual)	12	31	36	21	100	283	818
III (manual)	29	39	26	6	100	248	611
IV and V	36	34	20	9	100	242	897
Total*	**22**	**30**	**31**	**17**	**100**	**267**	3811
Document literacy							
I and II	10	19	38	33	100	298	1254
III (non-manual)	15	28	35	22	100	281	818
III (manual)	30	34	25	10	100	251	611
IV and V	37	30	24	9	100	241	897
Total*	**23**	**27**	**31**	**19**	**100**	**268**	3811
Quantitative literacy							
I and II	9	18	37	35	100	301	1254
III (non-manual)	17	28	37	18	100	277	818
III (manual)	28	36	26	11	100	254	611
IV and V	39	32	22	8	100	239	897
Total*	**23**	**28**	**30**	**19**	**100**	**267**	3811

* Informants who were members of the Armed Forces, who had never worked or who were in inadequately described occupations are not shown as separate categories but are included in the total figures.

Table A2.6a **Literacy level by social class (cumulative percentages)**

Social Class	Level 4/5	Level 3 or above	Level 2 or above	Level 1 or above
		Cumulative %		
Prose literacy				
I and II	28	69	90	100
III (non-manual)	21	57	88	100
III (manual)	6	32	71	100
IV and V	9	29	63	100
Total*	**17**	**48**	**78**	**100**
Document literacy				
I and II	33	71	90	100
III (non-manual)	22	57	85	100
III (manual)	10	35	69	100
IV and V	9	33	63	100
Total*	**19**	**50**	**77**	**100**
Quantitative literacy				
I and II	35	72	90	100
III (non-manual)	18	55	83	100
III (manual)	11	37	73	100
IV and V	8	30	62	100
Total*	**19**	**49**	**77**	**100**

* Informants who were members of the Armed Forces, who had never worked or who were in inadequately described occupations are not shown as separate categories but are included in the total figures.

Table A2. 7 **Literacy level by gross personal income from all sources**

Quintiles**	Level 1	Level 2	Level 3	Level 4/5	Total	Mean score	Base
	%	%	%	%	%		
Prose literacy							
Up to £2,704 pa	29	32	27	12	100	251	720
£2,705-£5,928 pa	33	32	25	10	100	246	779
£5,929-£10,400 pa	23	35	30	12	100	262	729
£10,401-£16,848 pa	16	30	36	18	100	277	675
£16,849 or more pa	5	22	41	32	100	302	755
Total*	**22**	**30**	**31**	**17**	**100**	**267**	3811
Document literacy							
Up to £2,704 pa	31	31	25	14	100	249	720
£2,705-£5,928 pa	39	27	26	8	100	240	779
£5,929-£10,400 pa	23	32	31	15	100	263	729
£10,401-£16,848 pa	15	29	34	22	100	280	675
£16,849 or more pa	5	17	39	39	100	311	755
Total*	**23**	**27**	**31**	**19**	**100**	**268**	3811
Quantitative literacy							
Up to £2,704 pa	34	32	24	9	100	242	720
£2,705-£5,928 pa	37	30	26	7	100	241	779
£5,929-£10,400 pa	24	31	32	13	100	264	729
£10,401-£16,848 pa	14	31	34	20	100	279	675
£16,849 or more pa	4	16	36	45	100	315	755
Total*	**23**	**28**	**30**	**19**	**100**	**268**	3811

* Informants who refused or did not know their income are not shown as separate categories but are included in the total figures.

** The quintile bands were calculated from the personal income data of individuals aged 16-65 from the General Household Survey.

Table A2.8 **Literacy level in England, Scotland and Wales by age and sex**

	Level 1	Level 2	Level 3	Level 4/5	Total	Mean score	Base
	%	%	%	%	%		
Prose literacy							
England							
Men							
16-25	18	31	34	18	100	270	168
26-45	16	27	35	22	100	276	523
Over 45	27	32	28	13	100	255	431
Total	**20**	**29**	**33**	**18**	**100**	**268**	**1122**
Women							
16-25	16	30	33	21	100	278	204
26-45	18	29	32	21	100	275	649
Over 45	32	33	27	8	100	248	497
Total	**22**	**31**	**30**	**17**	**100**	**266**	**1350**
All							
16-25	17	30	33	19	100	274	372
26-45	17	28	33	21	100	276	1172
Over 45	30	33	27	11	100	251	928
Total	**21**	**30**	**31**	**17**	**100**	**267**	**2472**
Scotland							
Men							
16-25	17	28	42	13	100	*	48
26-45	19	34	33	14	100	272	161
Over 45	36	28	25	11	100	253	113
Total	**24**	**31**	**32**	**13**	**100**	**265**	**322**
Women							
16-25	13	25	27	35	100	*	52
26-45	21	36	29	14	100	269	189
Over 45	29	31	31	8	100	254	141
Total	**22**	**32**	**29**	**16**	**100**	**268**	**382**
All							
16-25	15	26	35	24	100	279	100
26-45	20	35	31	14	100	270	350
Over 45	33	30	28	10	100	254	254
Total	**23**	**32**	**31**	**14**	**100**	**266**	**704**
Wales							
Men							
16-25	29	33	20	18	100	*	35
26-45	17	33	42	8	100	267	127
Over 45	30	39	26	5	100	243	124
Total	**25**	**35**	**32**	**8**	**100**	**256**	**286**
Women							
16-25	19	35	29	17	100	*	42
26-45	20	33	37	9	100	263	186
Over 45	30	32	32	6	100	252	121
Total	**23**	**33**	**34**	**9**	**100**	**261**	**349**
All							
16-25	25	34	24	18	100	265	77
26-45	19	33	39	9	100	265	313
Over 45	30	36	28	6	100	247	245
Total	**24**	**34**	**33**	**9**	**100**	**258**	**635**

Table A2.8 (continued) **Literacy level in England, Scotland and Wales by age and sex**

	Level 1	Level 2	Level 3	Level 4/5	Total	Mean score	Base
	%	%	%	%	%		
Document literacy							
England							
Men							
16-25	15	25	36	25	100	279	168
26-45	17	21	32	29	100	285	523
Over 45	26	29	27	17	100	260	431
Total	**19**	**25**	**31**	**24**	**100**	**275**	**1122**
Women							
16-25	21	28	33	18	100	272	204
26-45	21	27	32	20	100	271	649
Over 45	38	30	25	7	100	240	497
Total	**27**	**28**	**30**	**15**	**100**	**260**	**1350**
All							
16-25	18	26	35	21	100	276	372
26-45	19	24	32	25	100	278	1172
Over 45	32	30	26	12	100	250	928
Total	**23**	**26**	**31**	**20**	**100**	**268**	**2472**
Scotland							
Men							
16-25	14	32	32	22	100	*	48
26-45	18	28	31	23	100	279	161
Over 45	30	28	29	12	100	256	113
Total	**21**	**29**	**30**	**20**	**100**	**272**	**322**
Women							
16-25	16	27	31	26	100	*	52
26-45	22	35	27	16	100	266	189
Over 45	30	36	28	6	100	250	141
Total	**24**	**33**	**28**	**15**	**100**	**263**	**382**
All							
16-25	15	29	31	24	100	278	100
26-45	20	31	29	20	100	273	350
Over 45	30	32	28	9	100	253	254
Total	**22**	**31**	**29**	**17**	**100**	**267**	**704**
Wales							
Men							
16-25	23	29	23	26	100	*	35
26-45	18	29	36	17	100	276	127
Over 45	29	37	25	10	100	244	124
Total	**23**	**32**	**29**	**16**	**100**	**262**	**286**
Women							
16-25	19	32	27	22	100	*	42
26-45	26	29	35	11	100	258	186
Over 45	36	31	26	7	100	246	121
Total	**28**	**30**	**31**	**11**	**100**	**256**	**349**
All							
16-25	21	30	24	24	100	273	77
26-45	22	29	35	14	100	266	313
Over 45	32	34	25	8	100	244	245
Total	**26**	**31**	**30**	**13**	**100**	**258**	**635**

Table A2.8 (continued) **Literacy level in England, Scotland and Wales by age and sex**

	Level 1	Level 2	Level 3	Level 4/5	Total	Mean score	Base
	%	%	%	%	%		
Quantitative literacy							
England							
Men							
16-25	17	31	31	21	100	279	168
26-45	14	24	29	33	100	285	523
Over 45	22	27	30	21	100	260	431
Total	17	27	30	26	100	279	1122
Women							
16-25	28	26	34	12	100	272	204
26-45	23	27	33	16	100	271	649
Over 45	36	31	26	6	100	240	497
Total	29	28	31	12	100	256	1350
All							
16-25	22	29	32	16	100	265	372
26-45	19	26	31	25	100	278	1172
Over 45	29	29	28	13	100	255	928
Total	23	27	30	19	100	268	2472
Scotland							
Men							
16-25	18	28	39	15	100	*	48
26-45	17	23	35	25	100	279	161
Over 45	27	30	28	15	100	256	113
Total	21	26	34	20	100	274	322
Women							
16-25	24	29	34	13	100	*	52
26-45	27	36	26	11	100	266	189
Over 45	31	35	25	9	100	250	141
Total	28	34	27	11	100	258	382
All							
16-25	21	28	36	14	100	266	100
26-45	22	29	31	18	100	272	350
Over 45	29	32	27	12	100	258	254
Total	24	30	31	15	100	266	704
Wales							
Men							
16-25	28	30	24	18	100	*	35
26-45	16	26	42	16	100	276	127
Over 45	27	28	34	11	100	244	124
Total	23	28	35	14	100	266	286
Women							
16-25	24	36	30	9	100	*	42
26-45	26	33	31	10	100	258	186
Over 45	33	35	25	7	100	246	121
Total	28	34	29	9	100	254	349
All							
16-25	26	33	27	14	100	261	77
26-45	22	30	36	13	100	265	313
Over 45	30	31	30	9	100	252	245
Total	25	31	32	12	100	260	635

Table A2.9 **Literacy level by highest level of educational attainment in England, Scotland and Wales**

Eduaction level	Level 1	Level 2	Level 3	Level 4/5	Total	Mean score	Base
	%	%	%	%	%		
Prose literacy							
England							
Second level 1st stage or lower	30	35	26	9	100	248	1381
Second level 2nd stage	12	30	36	22	100	283	457
Third level	4	15	42	39	100	310	634
Total	**21**	**30**	**31**	**17**	**100**	**267**	**2472**
Scotland							
Second level 1st stage or lower	34	37	23	6	100	247	365
Second level 2nd stage	13	30	39	18	100	279	158
Third level	3	19	44	33	100	308	181
Total	**23**	**32**	**31**	**14**	**100**	**266**	**704**
Wales							
Second level 1st stage or lower	32	36	27	5	100	244	378
Second level 2nd stage	18	34	39	9	100	270	113
Third level	1	29	47	23	100	299	144
Total	**24**	**34**	**33**	**9**	**100**	**258**	**635**
Document literacy							
England							
Second level 1st stage or lower	32	30	27	11	100	248	1381
Second level 2nd stage	13	25	36	26	100	286	457
Third level	5	16	38	41	100	313	634
Total	**23**	**26**	**31**	**20**	**100**	**268**	**2472**
Scotland							
Second level 1st stage or lower	32	38	24	7	100	246	365
Second level 2nd stage	14	30	28	28	100	284	158
Third level	5	14	46	35	100	307	181
Total	**22**	**31**	**29**	**17**	**100**	**267**	**704**
Wales							
Second level 1st stage or lower	34	33	25	7	100	244	378
Second level 2nd stage	16	29	34	20	100	277	113
Third level	6	24	42	28	100	298	144
Total	**26**	**31**	**30**	**13**	**100**	**258**	**635**
Quantitative literacy							
England							
Second level 1st stage or lower	32	31	28	9	100	246	1381
Second level 2nd stage	12	27	35	25	100	286	457
Third level	5	16	33	47	100	316	634
Total	**23**	**27**	**30**	**19**	**100**	**268**	**2472**
Scotland							
Second level 1st stage or lower	33	36	24	6	100	246	365
Second level 2nd stage	18	23	39	20	100	280	158
Third level	5	20	40	35	100	308	181
Total	**24**	**30**	**31**	**15**	**100**	**266**	**704**
Wales							
Second level 1st stage or lower	34	35	26	6	100	244	378
Second level 2nd stage	16	27	43	14	100	274	113
Third level	5	22	44	29	100	302	144
Total	**25**	**31**	**32**	**12**	**100**	**260**	**635**

Table A2.10 **Literacy level by economic activity status in England, Scotland and Wales**

Economic activity status	Level 1	Level 2	Level 3	Level 4/5	Total	Mean score	Base
	%	%	%	%	%		
Prose literacy							
England							
Working	15	29	35	20	100	279	1691
Not working	36	32	22	10	100	239	781
Total	**21**	**30**	**31**	**17**	**100**	**267**	**2472**
Scotland							
Working	21	32	29	17	100	271	489
Not working	27	31	35	7	100	257	215
Total	**23**	**32**	**31**	**14**	**100**	**266**	**704**
Wales							
Working	17	37	37	9	100	267	435
Not working	39	28	25	8	100	239	200
Total	**24**	**34**	**33**	**9**	**100**	**258**	**635**
Document literacy							
England							
Working	16	25	35	24	100	282	1691
Not working	39	29	22	10	100	235	781
Total	**23**	**26**	**31**	**20**	**100**	**268**	**2472**
Scotland							
Working	20	29	31	20	100	274	489
Not working	29	36	24	11	100	253	215
Total	**22**	**31**	**29**	**17**	**100**	**267**	**704**
Wales							
Working	19	34	32	15	100	269	435
Not working	41	24	26	10	100	237	200
Total	**26**	**31**	**30**	**13**	**100**	**258**	**635**
Quantitative literacy							
England							
Working	16	27	33	24	100	282	1691
Not working	40	29	23	8	100	235	781
Total	**23**	**27**	**30**	**19**	**100**	**268**	**2472**
Scotland							
Working	22	29	32	18	100	272	489
Not working	30	32	28	10	100	253	215
Total	**24**	**30**	**31**	**15**	**100**	**266**	**704**
Wales							
Working	19	33	35	13	100	269	435
Not working	40	27	25	9	100	238	200
Total	**25**	**31**	**32**	**12**	**100**	**260**	**635**

Table A2.11 **Modelling literacy levels - results of multiple regression analysis**

Characteristic	Prose literacy coefficients	Document literacy	Quantitative literacy
Sex		[9]	[6]
Male	**	0.14	0.26
Female	**	*	*
Age-group	[3]	[3]	[4]
56-65	0.15	0.16	0.10
46-55	0.30	0.32	0.20
36-45	0.45	0.48	0.30
26-35	0.60	0.64	0.40
16-25	0.75	0.80	0.50
Highest level of educational attainment	[1]	[1]	[1]
Primary education or lower	0.15	0.15	0.17
Second level, 1st stage	0.30	0.30	0.34
Second level, 2nd stage	0.45	0.45	0.51
Third level, non-university	0.60	0.60	0.68
Third level, university	0.75	0.75	0.85
Social class	[2]	[2]	[3]
IV and V	0.17	0.17	0.18
IIIM (manual)	0.34	0.34	0.36
IIINM (non-manual)	0.51	0.51	0.54
I and II	0.68	0.68	0.72
Gross personal income	[8]	[4]	[2]
No income	0.06	0.09	0.09
Quintile 1 (upto £2,704 pa)	0.12	0.18	0.18
Quintile 2 (£2,705-£5,928 pa)	0.18	0.27	0.27
Quintile 3 (£5,929-£10,400 pa)	0.24	0.36	0.36
Quintile 4 (£10,401-£16,848 pa)	0.30	0.45	0.45
Quintile 5 (£16,849 pa or more)	0.36	0.54	0.54
Receipt of social security benefits (excluding pensions and child benefit)	[7]	[7]	[8]
Receives benefits	*	*	*
Does not receive benefits	0.18	0.23	0.23
Ethnic group	[9]	[10]	[10]
White	0.31	0.31	0.21
Non-white	*	*	*
Language first spoken	[5]	[6]	[7]
English	0.43	0.37	0.38
Other language	*	*	*
Frequency of reading books	[4]	[5]	[5]
At least once a week	0.36	0.34	0.26
Less often than once a week	*	*	*
Frequency of watching television	[6]	[8]	[9]
5 hours or more per day	*	*	*
less than 5 hours per day	0.28	0.26	0.27
Constant	-0.33	-0.47	-0.31
Percentage of variation explained by the model	35%	35%	36%

* Reference category for characteristic.
** Sex was not included as one of the independent variables for prose literacy.
The figures in brackets show the order in which the characteristics entered the model.

Annex tables

A3.1-A3.17

Table A3.1 **Literacy level by industry**
People who had worked in the 12 months prior to interview

	Level 1	Level 2	Level 3	Level 4/5	Total	Mean score	Base	% change in no. of employees 1983-96*
	%	%	%	%	%			
Prose literacy								
Agriculture/Mining/Supply of electricity, gas and water	22	37	30	11	100	261	112	-47
Manufacturing	16	33	33	18	100	275	620	-22
Construction	21	41	30	7	100	259	169	-23
Wholesale and retail	16	32	36	16	100	274	378	15
Hotel and restaurants	14	34	37	15	100	280	128	33
Transport/storage and telecommunications	31	26	30	13	100	259	153	-5
Finance	4	17	51	27	100	300	114	22
Real estate, renting, research, computer and other business	12	23	39	25	100	288	222	59
Public administration	6	20	35	39	100	305	172	-9
Education	12	22	35	31	100	293	256	16
Health and social work	18	26	35	21	100	278	315	35
Other community, social and personal services	19	34	26	21	100	273	143	24
Total	**16**	**30**	**35**	**20**	**100**	**278**	**2782**	**5**
Document literacy								
Agriculture/Mining/Supply of electricity, gas and water	18	29	39	15	100	272	112	-47
Manufacturing	15	29	34	22	100	280	620	-22
Construction	18	37	30	15	100	268	169	-23
Wholesale and retail	20	28	32	20	100	274	378	15
Hotel and restaurants	18	28	36	19	100	276	128	33
Transport/storage and telecommunications	29	28	29	14	100	258	153	-5
Finance	4	14	45	37	100	309	114	22
Real estate, renting, research, computer and other business	14	24	30	32	100	293	222	59
Public administration	6	19	35	40	100	310	172	-9
Education	14	19	34	33	100	295	256	16
Health and social work	18	26	39	17	100	273	315	35
Other community, social and personal services	27	23	32	19	100	269	143	24
Total	**17**	**26**	**34**	**23**	**100**	**280**	**2782**	**5**
Quantitative literacy								
Agriculture/Mining/Supply of electricity, gas and water	22	29	39	9	100	266	112	-47
Manufacturing	17	27	29	26	100	282	620	-22
Construction	16	36	34	14	100	272	169	-23
Wholesale and retail	21	29	34	17	100	271	378	15
Hotel and restaurants	20	38	33	9	100	264	128	33
Transport/storage and telecommunications	24	27	31	19	100	267	153	-5
Finance	4	14	48	35	100	311	114	22
Real estate, renting, research, computer and other business	12	27	28	33	100	294	222	59
Public administration	6	21	35	38	100	309	172	-9
Education	14	19	33	34	100	294	256	16
Health and social work	18	30	38	14	100	270	315	35
Other community, social and personal services	21	27	31	21	100	269	143	24
Total	**17**	**27**	**33**	**23**	**100**	**280**	**2782**	**5**

* The percentage change in the number of employees in employment in Great Britain by industrial sector has been calculated from *Labour Market Trends* April 1997, Table 1.2.

Table A3.2 **Literacy level by occupation and sex**
People who had worked in the 12 months prior to interview

	Level 1	Level 2	Level 3	Level 4/5	Total	Mean score	Base
	%	%	%	%	%		
Prose literacy							
Men							
Managers and administrators	8	29	39	24	100	289	*245*
Professional occupations	2	14	46	38	100	313	*199*
Associate professional and technical occupations	4	16	47	34	100	304	*132*
Clerical and secretarial occupations	16	30	30	24	100	277	*109*
Craft and related occupations	23	40	32	6	100	255	*261*
Personal and protective service occupations	12	19	45	23	100	292	*95*
Sales occupations	5	20	45	29	100	*	*59*
Plant and machine operators	31	40	24	5	100	248	*197*
Other occupations	38	31	24	7	100	241	*94*
Total	**16**	**29**	**35**	**19**	**100**	**276**	*1391*
Women							
Managers and administrators	8	26	38	27	100	294	*146*
Professional occupations	6	16	39	40	100	310	*157*
Associate professional and technical occupations	10	16	43	31	100	300	*150*
Clerical and secretarial occupations	9	28	41	22	100	290	*309*
Craft and related occupations	33	36	28	4	100	*	*38*
Personal and protective service occupations	21	34	31	14	100	268	*219*
Sales occupations	11	36	32	20	100	280	*161*
Plant and machine operators	37	44	11	9	100	237	*83*
Other occupations	33	40	22	5	100	247	*149*
Total	**16**	**30**	**34**	**21**	**100**	**279**	*1412*
All							
Managers and administrators	8	28	39	25	100	291	*391*
Professional occupations	3	15	43	39	100	312	*356*
Associate professional and technical occupations	7	16	45	33	100	302	*282*
Clerical and secretarial occupations	11	29	38	23	100	286	*418*
Craft and related occupations	24	39	31	6	100	255	*299*
Personal and protective service occupations	18	29	36	18	100	276	*314*
Sales occupations	10	31	36	23	100	285	*220*
Plant and machine operators	32	41	20	6	99	246	*280*
Other occupations	35	36	23	6	100	244	*243*
Total	**16**	**29**	**35**	**20**	**100**	**278**	*2803*
Document literacy							
Men							
Managers and administrators	7	21	39	32	100	300	*245*
Professional occupations	3	8	37	52	100	327	*199*
Associate professional and technical occupations	6	13	40	41	100	312	*132*
Clerical and secretarial occupations	18	21	30	31	100	285	*109*
Craft and related occupations	20	38	29	13	100	265	*261*
Personal and protective service occupations	10	17	41	32	100	300	*95*
Sales occupations	6	11	50	33	100	*	*59*
Plant and machine operators	25	39	27	10	100	257	*197*
Other occupations	33	31	26	10	100	243	*94*
Total	**15**	**25**	**34**	**26**	**100**	**285**	*1391*
Women							
Managers and administrators	12	30	38	21	100	286	*146*
Professional occupations	6	17	38	38	100	307	*157*
Associate professional and technical occupations	11	23	38	28	100	293	*150*
Clerical and secretarial occupations	11	28	38	23	100	287	*309*
Craft and related occupations	36	24	33	8	100	*	*38*
Personal and protective service occupations	25	31	36	9	100	260	*219*
Sales occupations	20	30	28	21	100	272	*161*
Plant and machine operators	41	41	12	6	100	232	*83*
Other occupations	37	26	33	4	100	239	*149*
Total	**19**	**28**	**34**	**18**	**100**	**273**	*1412*
All							
Managers and administrators	9	24	38	29	100	295	*391*
Professional occupations	4	12	38	47	100	319	*356*
Associate professional and technical occupations	8	17	40	35	100	303	*282*
Clerical and secretarial occupations	13	26	36	25	100	286	*418*
Craft and related occupations	21	37	30	12	100	263	*299*
Personal and protective service occupations	20	26	37	17	100	274	*314*
Sales occupations	16	25	35	24	100	282	*220*
Plant and machine operators	29	39	23	9	100	250	*280*
Other occupations	35	28	30	7	100	241	*243*
Total	**17**	**26**	**34**	**23**	**100**	**280**	*2803*

103

Table A3.2 (continued) **Literacy level by occupation and sex**
People who had worked in the 12 months prior to interview

	Level 1	Level 2	Level 3	Level 4/5	Total	Mean score	Base
	%	%	%	%	%		
Quantitative literacy							
Men							
Managers and administrators	7	15	42	36	100	306	245
Professional occupations	2	10	33	56	100	331	199
Associate professional and technical occupations	3	13	36	48	100	320	132
Clerical and secretarial occupations	17	23	29	30	100	286	109
Craft and related occupations	18	40	26	16	100	269	261
Personal and protective service occupations	10	27	37	25	100	292	95
Sales occupations	3	29	29	40	100	*	59
Plant and machine operators	21	39	28	12	100	263	197
Other occupations	32	37	21	10	100	244	94
Total	**13**	**27**	**31**	**29**	**100**	**289**	**1391**
Women							
Managers and administrators	10	28	40	22	100	289	146
Professional occupations	6	18	38	38	100	304	157
Associate professional and technical occupations	11	25	40	24	100	290	150
Clerical and secretarial occupations	13	27	42	18	100	281	309
Craft and related occupations	47	23	26	4	100	*	38
Personal and protective service occupations	29	36	31	5	100	253	219
Sales occupations	25	26	36	13	100	265	161
Plant and machine operators	42	38	15	5	100	229	83
Other occupations	41	33	25	2	100	235	149
Total	**22**	**29**	**35**	**15**	**100**	**269**	**1412**
All							
Managers and administrators	8	19	41	31	100	300	391
Professional occupations	3	13	35	49	100	320	356
Associate professional and technical occupations	7	19	38	37	100	306	282
Clerical and secretarial occupations	14	26	38	22	100	283	418
Craft and related occupations	21	38	26	14	100	265	299
Personal and protective service occupations	22	33	33	12	100	266	314
Sales occupations	19	27	34	21	100	277	220
Plant and machine operators	26	39	25	11	100	254	280
Other occupations	37	35	23	6	100	239	243
Total	**17**	**28**	**33**	**23**	**100**	**280**	**2803**

Table A3.3 **Literacy level by highest level of educational attainment and occupation**
People who had worked in the 12 months prior to interview and whose highest level of educational attainment was an upper or lower secondary education or lower

	Level 1	Level 2	Level 3	Level 4/5	Total	Mean score	Base
	%	%	%	%	%		
Prose literacy							
Lower secondary education or lower							
Managers and administrators	15	37	33	15	100	271	147
Professional and associate professional occupations	12	24	44	19	100	282	90
Clerical and secretarial occupations	14	33	37	16	100	277	268
Craft and related occupations	34	40	25	2	100	239	145
Personal and protective service occupations	24	29	36	11	100	267	197
Sales occupations	11	37	35	17	100	277	139
Plant and machine operatives	38	39	17	6	100	240	205
Other occupations	39	37	19	5	100	237	201
Total	**24**	**35**	**30**	**11**	**100**	**260**	**1392**
Upper secondary education							
Managers and administrators	0	31	50	18	100	297	85
Professional and associate professional occupations	4	11	51	34	100	308	71
Clerical and secretarial occupations	7	19	37	37	100	305	85
Craft and related occupations	16	42	33	10	100	266	117
Personal and protective service occupations	8	29	35	28	100	293	76
Sales occupations	6	23	38	33	100	*	60
Plant and machine operatives	16	50	27	6	100	*	64
Other occupations	20	28	41	11	100	*	31
Total	**9**	**30**	**39**	**22**	**100**	**287**	**589**
Document literacy							
Lower secondary education or lower							
Managers and administrators	17	31	39	13	100	274	147
Professional and associate professional occupations	16	26	35	23	100	284	90
Clerical and secretarial occupations	16	29	35	20	100	278	268
Craft and related occupations	29	37	26	7	100	247	145
Personal and protective service occupations	25	28	36	11	100	264	197
Sales occupations	19	30	31	20	100	274	139
Plant and machine operatives	33	40	21	7	100	243	205
Other occupations	39	28	27	6	100	234	201
Total	**25**	**31**	**31**	**13**	**100**	**261**	**1392**
Upper secondary education							
Managers and administrators	2	23	47	28	100	304	85
Professional and associate professional occupations	4	11	38	47	100	317	71
Clerical and secretarial occupations	8	17	37	38	100	303	85
Craft and related occupations	16	40	27	17	100	275	117
Personal and protective service occupations	11	20	42	28	100	291	76
Sales occupations	12	15	38	35	100	*	60
Plant and machine operatives	16	40	35	9	100	*	64
Other occupations	19	24	46	11	100	*	31
Total	**10**	**25**	**37**	**27**	**100**	**291**	**589**
Quantitative literacy							
Lower secondary education or lower							
Managers and administrators	13	26	43	18	100	280	147
Professional and associate professional occupations	13	30	36	21	100	284	90
Clerical and secretarial occupations	17	29	41	14	100	274	268
Craft and related occupations	31	38	23	8	100	249	145
Personal and protective service occupations	28	33	30	9	100	257	197
Sales occupations	22	31	32	14	100	267	139
Plant and machine operatives	30	40	23	7	100	246	205
Other occupations	41	34	20	4	100	232	201
Total	**25**	**33**	**31**	**11**	**100**	**260**	**1392**
Upper secondary education							
Managers and administrators	2	14	57	26	100	308	85
Professional and associate professional occupations	4	13	42	42	100	318	71
Clerical and secretarial occupations	10	20	29	42	100	300	85
Craft and related occupations	12	42	29	17	100	277	117
Personal and protective service occupations	12	31	41	16	100	281	76
Sales occupations	13	20	33	34	100	*	60
Plant and machine operatives	12	39	33	16	100	*	64
Other occupations	20	30	39	11	100	*	31
Total	**10**	**27**	**38**	**25**	**100**	**291**	**589**

105

Table A3.4 **Literacy level by whether works full time or part time and sex**
People who had worked in the 12 months prior to interview

	Level 1	Level 2	Level 3	Level 4/5	Total	Mean score	Base
	%	%	%	%	%		
Prose literacy							
Men							
Full time	17	29	35	19	100	276	*1226*
Part time	10	31	39	20	100	284	*97*
Total	**16**	**29**	**35**	**19**	**100**	**278**	*1323*
Women							
Full time	15	27	35	23	100	282	*708*
Part time	15	33	34	19	100	278	*607*
Total	**15**	**30**	**34**	**21**	**100**	**279**	*1315*
All							
Full time	16	29	35	20	100	278	*1934*
Part time	14	32	34	19	100	279	*704*
Total	**16**	**30**	**35**	**20**	**100**	**278**	*2638*
Document literacy							
Men							
Full time	15	26	33	26	100	285	*1226*
Part time	8	20	47	25	100	292	*97*
Total	**15**	**25**	**34**	**26**	**100**	**282**	*1323*
Women							
Full time	18	27	34	21	100	277	*708*
Part time	20	28	35	17	100	271	*607*
Total	**19**	**28**	**34**	**19**	**100**	**274**	*1315*
All							
Full time	16	26	33	25	100	282	*1934*
Part time	18	27	37	18	100	274	*704*
Total	**17**	**26**	**34**	**23**	**100**	**280**	*2638*
Quantitative literacy							
Men							
Full time	14	26	31	30	100	289	*1226*
Part time	9	31	40	21	100	288	*97*
Total	**13**	**26**	**31**	**29**	**100**	**283**	*1323*
Women							
Full time	20	27	36	17	100	272	*708*
Part time	23	30	34	14	100	267	*607*
Total	**21**	**28**	**35**	**15**	**100**	**271**	*1315*
All							
Full time	16	26	32	26	100	283	*1934*
Part time	20	30	35	15	100	271	*704*
Total	**17**	**27**	**33**	**23**	**100**	**280**	*2638*

Table A3.5 **Literacy level by frequency of engaging in several reading activities at work**

People who had worked in the 12 months prior to interview

		Level 1	Level 2	Level 3	Level 4/5	Total	Mean score	Base
		%	%	%	%	%		
Prose literacy								
Letters or memos	At least once a week	12	27	38	23	100	285	2056
	Less than once a week	26	36	26	12	100	257	701
Reports, articles, magazines or journals	At least once a week	10	26	40	24	100	290	1690
	Less than once a week	24	34	28	14	100	261	1067
Manuals or reference books, including catalogues	At least once a week	10	26	40	24	100	290	1527
	Less than once a week	22	34	29	15	100	265	1229
Diagrams	At least once a week	11	27	40	22	100	287	1058
	Less than once a week	19	31	32	19	100	273	1699
Bills, invoices, spreadsheets or budget tables	At least once a week	11	27	38	24	100	287	1266
	Less than once a week	21	31	31	17	100	270	1491
Directions or instructions for medicines, recipes or other products	At least once a week	14	30	37	19	100	280	830
	Less than once a week	17	29	34	20	100	277	1926
Use information from computers	At least once a week	8	23	42	27	100	295	1497
	Less than once a week	25	37	26	12	100	259	1260
Total		**16**	**29**	**35**	**20**	**100**	**278**	**2757**
Document literacy								
Letters or memos	At least once a week	13	24	36	27	100	289	2056
	Less than once a week	26	31	30	13	100	257	701
Reports, articles, magazines or journals	At least once a week	11	24	36	29	100	294	1690
	Less than once a week	25	29	31	14	100	261	1067
Manuals or reference books, including catalogues	At least once a week	11	22	37	29	100	294	1527
	Less than once a week	23	31	31	15	100	263	1229
Diagrams	At least once a week	11	23	38	28	100	294	1058
	Less than once a week	20	28	32	20	100	272	1699
Bills, invoices, spreadsheets or budget tables	At least once a week	12	24	36	27	100	290	1266
	Less than once a week	21	28	33	19	100	272	1491
Directions or instructions for medicines, recipes or other products	At least once a week	16	23	39	22	100	285	830
	Less than once a week	17	27	33	23	100	278	1926
Use information from computers	At least once a week	9	21	39	32	100	299	1497
	Less than once a week	26	32	29	13	100	258	1260
Total		**17**	**26**	**34**	**23**	**100**	**280**	**2757**
Quantitative literacy								
Letters or memos	At least once a week	13	25	36	27	100	289	2056
	Less than once a week	28	33	27	12	100	255	701
Reports, articles, magazines or journals	At least once a week	11	24	36	29	100	294	1690
	Less than once a week	25	32	29	13	100	260	1067
Manuals or reference books, including catalogues	At least once a week	10	24	36	30	100	294	1527
	Less than once a week	24	32	30	14	100	263	1229
Diagrams	At least once a week	10	23	36	31	100	296	1058
	Less than once a week	21	30	31	18	100	270	1699
Bills, invoices, spreadsheets or budget tables	At least once a week	13	23	35	28	100	291	1266
	Less than once a week	20	31	31	18	100	270	1491
Directions or instructions for medicines, recipes or other products	At least once a week	16	27	35	22	100	282	830
	Less than once a week	17	27	32	23	100	279	1926
Use information from computers	At least once a week	9	21	38	32	100	299	1497
	Less than once a week	26	35	27	12	100	258	1260
Total		**17**	**27**	**33**	**23**	**100**	**280**	**2757**

Table A3.6 **Percentage of respondents at each literacy level who reported engaging in each of several reading activities at work at least once a week**

People who had worked in the 12 months prior to interview

	Letters or memos	Reports, articles, magazines or journals	Manuals or reference books, including catalogues	Diagrams	Bills, invoices, spreadsheets or budget tables	Directions or instructions for medicines, recipes or other products	Use information from computers	Base
	%	%	%	%	%	%	%	
Prose literacy								
Level 1	56	37	35	26	31	26	28	434
Level 2	68	52	48	34	43	30	42	815
Level 3	80	67	62	43	52	31	65	960
Level 4/5	84	72	65	41	56	28	72	548
Total	**73**	**59**	**54**	**37**	**47**	**30**	**54**	2757
Document literacy								
Level 1	58	39	36	24	35	28	29	470
Level 2	68	54	46	33	43	26	42	736
Level 3	77	62	59	41	50	33	60	932
Level 4/5	85	74	70	46	56	29	74	619
Total	**73**	**59**	**54**	**37**	**47**	**30**	**54**	2757
Quantitative literacy								
Level 1	55	38	34	22	35	28	28	477
Level 2	68	51	47	32	40	30	41	744
Level 3	79	64	59	40	50	31	62	914
Level 4/5	86	76	71	50	59	28	75	622
Total	**73**	**59**	**54**	**37**	**47**	**30**	**54**	2757

Table A3.7 **Literacy level by frequency of engaging in several writing and mathematical activities at work**

People who had worked in the 12 months prior to interview

		Level 1	Level 2	Level 3	Level 4/5	Total	Mean score	Base
		%	%	%	%	%		
Writing tasks								
Prose literacy								
Letters or memos	At least once a week	11	24	40	26	100	291	*1586*
	Less than once a week	22	36	28	13	100	262	*1171*
Forms or things such as bills, invoices or budget tables	At least once a week	11	27	40	22	100	286	*1369*
	Less than once a week	21	32	29	18	100	270	*1388*
Reports or articles	At least once a week	11	25	40	24	100	290	*1056*
	Less than once a week	19	32	32	17	100	271	*1701*
Estimates or technical specifications	At least once a week	11	28	43	19	100	284	*573*
	Less than once a week	17	30	33	20	100	277	*2183*
Total		**16**	**29**	**35**	**20**	**100**	**278**	*2757*
Document literacy								
Letters or memos	At least once a week	12	21	38	30	100	295	*1586*
	Less than once a week	23	33	30	14	100	262	*1171*
Forms or things such as bills, invoices or budget tables	At least once a week	11	25	37	27	100	289	*1369*
	Less than once a week	22	27	32	19	100	271	*1388*
Reports or articles	At least once a week	11	21	37	30	100	295	*1056*
	Less than once a week	20	29	33	19	100	272	*1701*
Estimates or technical specifications	At least once a week	10	25	36	29	100	292	*573*
	Less than once a week	18	26	34	21	100	277	*2183*
Total		**17**	**26**	**34**	**23**	**100**	**280**	*2757*
Quantitative literacy								
Letters or memos	At least once a week	11	22	36	31	100	295	*1586*
	Less than once a week	24	34	29	13	100	261	*1171*
Forms or things such as bills, invoices or budget tables	At least once a week	12	24	37	27	100	290	*1369*
	Less than once a week	22	31	30	18	100	269	*1388*
Reports or articles	At least once a week	11	23	34	32	100	296	*1056*
	Less than once a week	20	30	33	17	100	271	*1701*
Estimates or technical specifications	At least once a week	10	24	36	31	100	296	*573*
	Less than once a week	19	28	33	21	100	276	*2183*
Total		**17**	**27**	**33**	**23**	**100**	**280**	*2757*
Mathematical tasks								
Prose literacy								
Measure or estimate the size or weight of objects	At least once a week	15	34	35	16	100	274	*1199*
	Less than once a week	16	26	34	23	100	281	*1558*
Calculate prices, costs or budgets	At least once a week	10	27	40	22	100	287	*1198*
	Less than once a week	20	32	30	18	100	270	*1559*
Total		**16**	**29**	**35**	**20**	**100**	**278**	*2757*
Document literacy								
Measure or estimate the size or weight of objects	At least once a week	16	27	36	21	100	281	*1199*
	Less than once a week	18	25	33	25	100	280	*1558*
Calculate prices, costs or budgets	At least once a week	11	23	38	27	100	291	*1198*
	Less than once a week	21	28	32	19	100	272	*1559*
Total		**17**	**26**	**34**	**23**	**100**	**280**	*2757*
Quantitative literacy								
Measure or estimate the size or weight of objects	At least once a week	14	30	34	22	100	282	*1199*
	Less than once a week	19	25	32	24	100	279	*1558*
Calculate prices, costs or budgets	At least once a week	12	23	38	28	100	292	*1198*
	Less than once a week	21	31	29	18	100	270	*1559*
Total		**17**	**27**	**33**	**23**	**100**	**280**	*2757*

109

Table A3.8 **Percentage of respondents within a level who reported engaging in each of several workplace writing and mathematical tasks at least once a week**

People who had worked in the 12 months prior to interview

	Writing tasks				Mathematical tasks		
	Letters or memos	Forms or things such as bills, invoices or budgets	Reports or articles	Estimates or technical specifications	Measure or estimate the size or weight of objects	Calculate prices, costs or budgets	*Base*
	%	%	%	%	%	%	
Prose literacy							
Level 1	38	35	25	14	42	30	*434*
Level 2	45	46	31	19	50	41	*815*
Level 3	63	59	42	25	44	52	*960*
Level 4/5	71	56	44	19	34	50	*548*
Total	55	51	36	20	43	45	*2757*
Document literacy							
Level 1	39	35	25	12	40	31	*470*
Level 2	44	49	30	19	45	40	*736*
Level 3	61	54	39	21	46	49	*932*
Level 4/5	72	59	48	26	39	54	*619*
Total	55	51	36	20	43	45	*2757*
Quantitative literacy							
Level 1	36	35	24	11	36	31	*477*
Level 2	44	45	30	18	48	37	*744*
Level 3	61	56	38	22	44	51	*914*
Level 4/5	75	61	52	27	41	56	*622*
Total	55	51	36	20	43	45	*2757*

Table A3.9 **Self-assessment of reading skills for main job by literacy level**

People who had worked in the 12 months prior to interview

	Excellent	Good	Moderate	Poor	No opinion	Total	*Base*
	%	%	%	%	%	%	
Prose literacy							
Level 1	13	41	30	10	6	100	*434*
Level 2	26	53	18	2	2	100	*816*
Level 3	46	45	6	0	2	100	*960*
Level 4/5	68	28	3	-	1	100	*548*
Total	39	44	13	2	2	100	*2758*
Document literacy							
Level 1	16	41	28	10	5	100	*471*
Level 2	27	52	17	1	2	100	*736*
Level 3	46	44	7	1	2	100	*932*
Level 4/5	61	35	4	-	1	100	*619*
Total	39	44	13	2	2	100	*2758*
Quantitative literacy							
Level 1	17	40	27	10	5	100	*478*
Level 2	28	54	15	1	2	100	*744*
Level 3	45	44	8	1	2	100	*914*
Level 4/5	61	33	5	-	1	100	*622*
Total	39	44	13	2	2	100	*2758*

Table A3.10 **Self-assessment of writing skills for main job by literacy level**
People who had worked in the 12 months prior to interview

	Excellent	Good	Moderate	Poor	No opinion	Total	Base
	%	%	%	%	%	%	
Prose literacy							
Level 1	10	36	35	13	5	100	434
Level 2	17	51	27	3	2	100	814
Level 3	33	51	12	1	2	100	960
Level 4/5	51	41	6	0	2	100	548
Total	28	47	19	4	2	100	2756
Document literacy							
Level 1	12	39	32	12	5	100	469
Level 2	18	50	26	4	2	100	736
Level 3	34	49	14	1	2	100	932
Level 4/5	44	45	9	1	1	100	619
Total	28	47	19	4	2	100	2756
Quantitative literacy							
Level 1	13	38	33	12	5	100	477
Level 2	21	50	23	3	2	100	743
Level 3	31	51	15	2	2	100	914
Level 4/5	46	43	10	1	1	100	622
Total	28	47	19	4	2	100	2756

Table A3.11 **Self-assessment of mathematics skills for main job by literacy level**
People who had worked in the 12 months prior to interview

	Excellent	Good	Moderate	Poor	No opinion	Total	Base
	%	%	%	%	%	%	
Prose literacy							
Level 1	12	38	36	8	6	100	434
Level 2	18	48	27	4	3	100	815
Level 3	31	48	16	2	3	100	960
Level 4/5	37	48	10	2	4	100	548
Total	25	46	21	4	4	100	2757
Document literacy							
Level 1	10	38	37	9	5	100	470
Level 2	15	48	27	5	4	100	736
Level 3	28	52	17	1	2	100	932
Level 4/5	43	43	9	2	3	100	619
Total	25	46	21	4	4	100	2757
Quantitative literacy							
Level 1	8	32	42	12	6	100	477
Level 2	15	49	28	4	3	100	744
Level 3	28	52	14	1	4	100	914
Level 4/5	45	46	7	1	2	100	622
Total	25	46	21	4	4	100	2757

Table A3.12 **Whether reading skills were limiting job opportunities by literacy level**

People who had worked in the 12 months prior to interview

	Greatly limiting	Somewhat limiting	Not at all limiting	Total	Base
	%	%	%	%	
Prose literacy					
Level 1	8	20	72	100	431
Level 2	2	13	85	100	810
Level 3	1	6	94	100	959
Level 4/5	0	2	97	100	548
Total	2	9	88	100	2748
Document literacy					
Level 1	7	20	73	100	466
Level 2	2	13	85	100	733
Level 3	1	6	93	100	930
Level 4/5	0	4	95	100	619
Total	2	9	88	100	2748
Quantitative literacy					
Level 1	8	21	71	100	472
Level 2	1	10	89	100	742
Level 3	1	7	92	100	912
Level 4/5	1	4	95	100	622
Total	2	9	88	100	2748

Table A3.13 **Whether writing skills were limiting job opportunities by literacy level**

People who had worked in the 12 months prior to interview

	Greatly limiting	Somewhat limiting	Not at all limiting	Total	Base
	%	%	%	%	
Prose literacy					
Level 1	8	29	63	100	430
Level 2	2	19	80	100	813
Level 3	1	8	91	100	957
Level 4/5	-	2	98	100	548
Total	2	13	85	100	2748
Document literacy					
Level 1	8	27	65	100	467
Level 2	1	18	81	100	733
Level 3	1	8	91	100	929
Level 4/5	0	5	95	100	619
Total	2	13	85	100	2748
Quantitative literacy					
Level 1	7	28	65	100	475
Level 2	1	15	84	100	741
Level 3	1	10	89	100	910
Level 4/5	1	5	95	100	622
Total	2	13	85	100	2748

Table A3.14 **Whether mathematics skills were limiting job opportunities by literacy level**

People who had worked in the 12 months prior to interview

	Greatly limiting	Somewhat limiting	Not at all limiting	Total	Base
	%	%	%	%	
Prose literacy					
Level 1	8	25	67	100	431
Level 2	2	16	81	100	812
Level 3	1	11	88	100	960
Level 4/5	0	5	94	100	547
Total	2	14	84	100	2750
Document literacy					
Level 1	10	25	66	100	467
Level 2	2	20	78	100	735
Level 3	0	9	90	100	930
Level 4/5	0	5	95	100	618
Total	2	14	84	100	2750
Quantitative literacy					
Level 1	10	30	60	100	476
Level 2	2	16	82	100	740
Level 3	0	9	91	100	913
Level 4/5	0	5	95	100	621
Total	2	14	84	100	2750

Table A3.15 **Percentage who participated in adult education or training during the 12 months prior to interview by occupation and sex**

People who had worked in the 12 months prior to interview

	Males	Females	All
	Percentage who participated in adult education or training		
Managers and administrators	54	71	60
Professional occupations	78	81	79
Associate professional and technical occupations	61	85	71
Clerical and secretarial occupations	65	63	63
Craft and related occupations	40	28	38
Personal and protective service occupations	72	54	61
Sales occupations	74	58	63
Plant and machine operatives	43	30	40
Other occupations	41	36	39
Total	**55**	**59**	**57**
Base			
Managers and administrators	*245*	*146*	*391*
Professional occupations	*199*	*157*	*356*
Associate professional and technical occupations	*132*	*150*	*282*
Clerical and secretarial occupations	*109*	*309*	*418*
Craft and related occupations	*261*	*38*	*299*
Personal and protective service occupations	*95*	*219*	*314*
Sales occupations	*59*	*161*	*220*
Plant and machine operatives	*197*	*83*	*280*
Other occupations	*94*	*149*	*243*
Total	*1391*	*1412*	*2803*

Table A3.16 **Literacy level by annual gross income from employment**

Employed full-time

Wage income quintiles*	Level 1	Level 2	Level 3	Level 4/5	Total	Mean score	*Base*
	%	%	%	%	%		
Prose literacy							
Up to £4,628 pa	35	27	32	5	100	248	*65*
£4,629 to £8,996 pa	29	31	30	9	100	254	*222*
£8,997 to £13,000 pa	18	36	33	13	100	270	*412*
£13,001 to £19,188 pa	13	31	36	20	100	279	*482*
£19,189 pa or more	4	21	41	34	100	304	*558*
Total	**15**	**29**	**35**	**20**	**100**	**280**	*1739*
Document literacy							
Up to £4,628 pa	39	29	22	11	100	243	*65*
£4,629 to £8,996 pa	27	33	27	14	100	256	*222*
£8,997 to £13,000 pa	17	32	32	20	100	276	*412*
£13,001 to £19,188 pa	14	29	34	23	100	284	*482*
£19,189 pa or more	5	15	40	41	100	314	*558*
Total	**15**	**26**	**34**	**26**	**100**	**285**	*1739*
Quantitative literacy							
Up to £4,628 pa	38	35	23	4	100	235	*65*
£4,629 to £8,996 pa	27	33	28	11	100	255	*222*
£8,997 to £13,000 pa	18	32	33	18	100	274	*412*
£13,001 to £19,188 pa	12	29	34	25	100	285	*482*
£19,189 pa or more	4	13	36	47	100	319	*558*
Total	**14**	**26**	**33**	**27**	**100**	**286**	*1739*

*The quintile bands were calculated from the personal income data of individuals in employment aged 16-65 from the 1994 General Household Survey

Table A3.17 **Annual gross income from employment by age-group and literacy level**
Employed full-time

	Wage income quintiles*					Total	Base
	Up to £4,628 pa	£4,629 to £8,996 pa	£8,997 to £13,000 pa	£13,001 to £19,188 pa	£19,189 pa or more		
	%	%	%	%	%	%	
Prose literacy							
Age 16-25							
Level 1	19	48	27	6	-	100	32
Level 2	9	27	48	16	0	100	74
Level 3	5	39	30	22	4	100	79
Level 4/5	5	22	31	32	9	100	48
Total	8	33	36	19	3	100	233
Age 26-45							
Level 1	9	24	30	27	10	100	113
Level 2	2	13	25	32	28	100	255
Level 3	3	8	20	27	41	100	360
Level 4/5	-	4	15	25	57	100	250
Total	3	10	22	28	38	100	978
Age 46-65							
Level 1	5	27	31	26	11	100	106
Level 2	3	15	29	26	26	100	166
Level 3	3	4	23	25	45	100	178
Level 4/5	1	6	10	20	63	100	78
Total	3	12	25	25	35	100	528
All							
Level 1	9	29	30	23	9	100	251
Level 2	4	16	31	27	22	100	495
Level 3	3	12	23	26	36	100	617
Level 4/5	1	7	16	25	51	100	376
Total	4	15	25	26	31	100	1739
Document literacy							
Age 16-25							
Level 1	26	38	31	5	-	100	34
Level 2	9	37	39	16	0	100	65
Level 3	3	31	38	25	2	100	84
Level 4/5	5	30	30	24	11	100	50
Total	8	33	36	19	3	100	233
Age 26-45							
Level 1	8	22	29	27	14	100	120
Level 2	2	16	25	35	22	100	221
Level 3	3	7	21	27	43	100	325
Level 4/5	1	5	17	24	53	100	312
Total	3	10	22	28	38	100	978
Age 46-65							
Level 1	6	29	27	29	9	100	96
Level 2	5	12	34	27	23	100	160
Level 3	1	9	19	24	46	100	180
Level 4/5	1	1	18	18	62	100	92
Total	3	12	25	25	35	100	528
All							
Level 1	10	27	29	24	10	100	250
Level 2	4	19	30	29	18	100	446
Level 3	2	12	24	26	36	100	589
Level 4/5	2	8	19	23	49	100	454
Total	4	15	25	26	31	100	1739
Quantitative literacy							
Age 16-25							
Level 1	18	38	39	5	-	100	41
Level 2	12	35	41	12	0	100	67
Level 3	4	35	33	26	3	100	77
Level 4/5	1	24	28	35	12	100	48
Total	8	33	36	19	3	100	233
Age 26-45							
Level 1	9	22	29	29	11	100	119
Level 2	3	15	26	34	22	100	225
Level 3	2	8	22	28	39	100	313
Level 4/5	0	4	16	22	57	100	321
Total	3	10	22	28	38	100	978
Age 46-65							
Level 1	7	32	26	22	13	100	91
Level 2	3	15	33	32	18	100	140
Level 3	3	7	25	23	42	100	186
Level 4/5	1	3	12	20	63	100	111
Total	3	12	25	25	35	100	528
All							
Level 1	10	28	30	22	9	100	251
Level 2	5	19	31	29	16	100	432
Level 3	3	13	25	26	34	100	576
Level 4/5	1	6	16	23	53	100	480
Total	4	15	25	26	31	100	1739

*The quintile bands were calculated from the personal income data of individuals in employment aged 16-65 from the 1994 General Household Survey

Annex tables
A4.1-A4.16

Table A4.1 Literacy level by frequency of reading newspapers or magazines in daily life

	Level 1	Level 2	Level 3	Level 4/5	Total	Mean score	Base
	%	%	%	%	%		
Prose literacy							
Daily	21	32	32	15	100	268	2735
Weekly	19	26	31	24	100	276	819
Monthly	21	34	37	9	100	253	106
Several times a year	28	40	25	7	100	*	56
Never	71	15	9	4	100	151	88
Total	**22**	**30**	**31**	**17**	**100**	**267**	**3804**
Document literacy							
Daily	23	28	31	19	100	270	2735
Weekly	19	24	31	25	100	276	819
Monthly	25	32	34	8	100	248	106
Several times a year	28	25	32	15	100	*	56
Never	71	14	13	2	100	144	88
Total	**23**	**27**	**31**	**19**	**100**	**268**	**3804**
Quantitative literacy							
Daily	22	29	30	18	100	270	2735
Weekly	21	24	32	23	100	274	819
Monthly	26	33	31	11	100	252	106
Several times a year	35	25	32	9	100	*	56
Never	71	9	18	1	100	155	88
Total	**23**	**28**	**30**	**19**	**100**	**267**	**3804**

Table A4.2 Literacy level by frequency of reading books in daily life

	Level 1	Level 2	Level 3	Level 4/5	Total	Mean score	Base
	%	%	%	%	%		
Prose literacy							
Daily	11	27	37	25	100	288	1329
Weekly	14	30	33	24	100	283	615
Monthly	20	35	33	12	100	266	472
Several times a year	22	33	33	13	100	265	755
Never	49	31	16	4	100	216	631
Total	**22**	**30**	**31**	**17**	**100**	**267**	**3802**
Document literacy							
Daily	12	25	36	27	100	288	1329
Weekly	17	25	32	25	100	283	615
Monthly	23	33	29	15	100	265	472
Several times a year	22	29	31	18	100	270	755
Never	49	26	19	6	100	216	631
Total	**23**	**27**	**31**	**19**	**100**	**268**	**3802**
Quantitative literacy							
Daily	13	26	37	24	100	285	1329
Weekly	19	23	34	24	100	283	615
Monthly	25	34	26	15	100	264	472
Several times a year	22	31	28	19	100	269	755
Never	45	28	21	6	100	222	631
Total	**23**	**28**	**30**	**19**	**100**	**267**	**3802**

Table A4.3 Literacy level by frequency of writing letters of more than a page in daily life

	Level 1	Level 2	Level 3	Level 4/5	Total	Mean score	Base
	%	%	%	%	%		
Prose literacy							
Daily	10	32	34	24	100	285	149
Weekly	12	24	38	26	100	288	700
Monthly	16	29	34	21	100	279	829
Several times a year	19	33	33	15	100	270	1200
Never	39	32	22	7	100	233	927
Total	**22**	**30**	**31**	**17**	**100**	**267**	**3805**
Document literacy							
Daily	17	25	28	30	100	284	149
Weekly	14	25	36	25	100	285	700
Monthly	16	27	33	23	100	282	829
Several times a year	20	28	32	19	100	273	1200
Never	40	28	22	10	100	233	927
Total	**23**	**27**	**31**	**19**	**100**	**268**	**3805**
Quantitative literacy							
Daily	19	25	25	31	100	280	149
Weekly	15	24	40	21	100	282	700
Monthly	16	28	34	23	100	281	829
Several times a year	21	30	31	18	100	271	1200
Never	39	28	21	13	100	237	927
Total	**23**	**28**	**30**	**19**	**100**	**267**	**3805**

Table A4.4 Literacy level by whether household has more than 25 books

	Level 1	Level 2	Level 3	Level 4/5	Total	Mean score	Base
	%	%	%	%	%		
Prose literacy							
More than 25 books	16	30	34	19	100	277	3130
25 books or less	51	30	16	3	100	215	676
Total	**22**	**30**	**31**	**17**	**100**	**267**	**3806**
Document literacy							
More than 25 books	17	27	34	22	100	279	3130
25 books or less	53	27	14	5	100	211	676
Total	**23**	**27**	**31**	**19**	**100**	**268**	**3806**
Quantitative literacy							
More than 25 books	17	28	33	21	100	278	3130
25 books or less	53	26	17	4	100	213	676
Total	**23**	**28**	**30**	**19**	**100**	**267**	**3806**

Table A4.5 Literacy level by whether household has a dictionary

	Level 1	Level 2	Level 3	Level 4/5	Total	Mean score	Base
	%	%	%	%	%		
Prose literacy							
Yes	19	30	33	17	100	271	3477
No	53	28	13	6	100	211	325
Total	**22**	**30**	**31**	**17**	**100**	**267**	**3802**
Document literacy							
Yes	21	27	32	20	100	272	3477
No	55	27	12	6	100	209	325
Total	**23**	**27**	**31**	**19**	**100**	**268**	**3802**
Quantitative literacy							
Yes	21	28	32	19	100	272	3477
No	57	25	12	6	100	206	325
Total	**23**	**28**	**30**	**19**	**100**	**267**	**3802**

Table A4.6 Literacy level by whether household has a daily newspaper

	Level 1	Level 2	Level 3	Level 4/5	Total	Mean score	Base
	%	%	%	%	%		
Prose literacy							
Yes	21	31	32	16	100	267	2629
No	22	29	30	19	100	266	1177
Total	22	30	31	17	100	267	3806
Document literacy							
Yes	22	28	30	19	100	268	2629
No	25	24	31	20	100	265	1177
Total	23	27	30	19	100	268	3806
Quantitative literacy							
Yes	22	30	29	19	100	268	2629
No	26	23	33	18	100	265	1177
Total	23	28	30	19	100	267	3806

Table A4.7 Literacy level by frequency of using a public library

	Level 1	Level 2	Level 3	Level 4/5	Total	Mean score	Base
	%	%	%	%	%		
Prose literacy							
Daily	5	9	68	18	100	*	41
Weekly	15	27	35	22	100	280	435
Monthly	10	27	36	26	100	290	688
Several times a year	13	29	38	20	100	283	987
Never	33	34	23	10	100	243	1655
Total	22	30	31	17	100	267	3806
Document literacy							
Daily	6	8	45	41	100	*	41
Weekly	18	25	36	20	100	277	435
Monthly	11	25	36	28	100	292	688
Several times a year	15	25	35	25	100	285	987
Never	35	30	24	11	100	244	1655
Total	23	27	31	19	100	268	3806
Quantitative literacy							
Daily	6	27	49	18	100	*	41
Weekly	17	25	40	18	100	278	435
Monthly	14	25	35	26	100	286	688
Several times a year	16	27	33	24	100	282	987
Never	33	30	24	13	100	247	1655
Total	23	28	30	19	100	267	3806

Table A4.8 **Literacy level by amount of time spent watching television**

	Level 1	Level 2	Level 3	Level 4/5	Total	Mean score	Base
	%	%	%	%	%		
Prose literacy							
Not on a daily basis	23	24	31	21	100	269	167
1 hour or less per day	12	23	35	30	100	288	341
1 to 2 hours per day	16	27	35	22	100	278	966
More than 2 hours but less than 5	21	32	33	14	100	266	1801
5 or more hours per day	42	37	16	5	100	229	503
Do not have a television	[6]	[10]	[7]	[2]	[25]	*	25
Total	**22**	**30**	**31**	**17**	**100**	**267**	**3803**
Document literacy							
Not on a daily basis	23	26	29	23	100	269	167
1 hour or less per day	16	18	36	30	100	288	341
1 to 2 hours per day	17	24	33	26	100	281	966
More than 2 hours but less than 5	22	29	33	16	100	267	1801
5 or more hours per day	45	33	14	7	100	227	503
Do not have a television	[7]	[10]	[6]	[2]	[25]	*	25
Total	**23**	**27**	**31**	**19**	**100**	**268**	**3803**
Quantitative literacy							
Not on a daily basis	24	29	30	17	100	265	167
1 hour or less per day	15	20	33	31	100	289	341
1 to 2 hours per day	17	24	33	26	100	282	966
More than 2 hours but less than 5	22	30	32	16	100	267	1801
5 or more hours per day	47	31	16	5	100	225	503
Do not have a television	[7]	[7]	[7]	[4]	[25]	*	25
Total	**23**	**28**	**30**	**19**	**100**	**267**	**3803**

Table A4.9 **Amount of time spent watching television by frequency of reading books**

Amount of time spent watching television	Frequency of reading books					Total	Base
	Daily	Weekly	Monthly	Several times a year	Never		
	%	%	%	%	%	%	
Not on a daily basis	36	17	11	18	18	100	165
1 hour or less per day	36	24	12	18	11	100	341
1-2 hours per day	36	17	13	21	13	100	966
More than 2 hours but less than 5	30	15	14	23	18	100	1800
5 or more hours per day	23	10	11	23	32	100	502

Table A4.10 **Literacy level by how often respondent follows current events**

	Level 1	Level 2	Level 3	Level 4/5	Total	Mean score	Base
	%	%	%	%	%		
Prose literacy							
Most of the time	14	28	36	22	100	282	1899
Some of the time	19	31	34	16	100	270	1113
Only now and then	37	35	21	7	100	243	508
Hardly at all	51	31	13	4	100	206	282
Total	**22**	**30**	**31**	**17**	**100**	**267**	**3802**
Document literacy							
Most of the time	16	26	33	25	100	283	1899
Some of the time	20	26	35	18	100	270	1113
Only now and then	37	32	22	10	100	245	508
Hardly at all	54	26	17	3	100	203	282
Total	**23**	**27**	**31**	**19**	**100**	**268**	**3802**
Quantitative literacy							
Most of the time	15	26	34	26	100	286	1899
Some of the time	21	29	34	17	100	270	1113
Only now and then	39	35	20	7	100	237	508
Hardly at all	57	24	16	3	100	200	282
Total	**23**	**28**	**30**	**19**	**100**	**267**	**3802**

Table A4.11 **Literacy level by frequency of using a personal computer at home**

	Level 1	Level 2	Level 3	Level 4/5	Total	Mean score	Base
	%	%	%	%	%		
Prose literacy							
Daily	6	24	47	23	100	292	423
Weekly	6	19	40	35	100	303	411
Monthly	10	24	36	31	100	295	188
Several times a year	12	26	35	27	100	288	202
Never	29	34	27	10	100	252	2581
Total	**22**	**30**	**31**	**17**	**100**	**267**	**3805**
Document literacy							
Daily	8	18	43	31	100	299	423
Weekly	7	18	36	40	100	308	411
Monthly	9	22	41	29	100	297	188
Several times a year	13	22	32	32	100	288	202
Never	31	31	26	12	100	251	2581
Total	**23**	**27**	**31**	**19**	**100**	**268**	**3805**
Quantitative literacy							
Daily	8	21	41	30	100	298	423
Weekly	6	21	36	37	100	304	411
Monthly	11	19	42	28	100	295	188
Several times a year	16	26	30	28	100	283	202
Never	30	31	27	12	100	252	2581
Total	**23**	**28**	**30**	**19**	**100**	**267**	**3805**

Table A4.12 **Self-assessment of reading, writing and mathematics skills in daily life for all respondents**

	Excellent	Good	Moderate	Poor	No opinion	Total	Base
	%	%	%	%	%	%	
Reading skills	46	40	11	3	1	100	3805
Writing skills	33	45	17	5	1	100	3805
Mathematics skills	25	45	23	6	0	100	3805

Table A4.13 **Literacy level by self-assessment of reading skills for daily life**

	Level 1	Level 2	Level 3	Level 4/5	Total	Mean score	Base
	%	%	%	%	%		
Prose literacy							
Excellent	9	25	40	26	100	293	1834
Good	23	37	29	11	100	262	1487
Moderate	56	33	9	2	100	213	392
Poor	86	8	6	-	100	128	77
No opinion	[12]	[3]	-	-	[15]	*	15
Total	**22**	**30**	**31**	**17**	**100**	**267**	**3805**
Document literacy							
Excellent	10	24	38	28	100	294	1834
Good	24	32	28	15	100	264	1487
Moderate	56	28	12	3	100	216	392
Poor	87	5	8	-	100	114	77
No opinion	[13]	-	[2]	-	[15]	*	15
Total	**23**	**27**	**31**	**19**	**100**	**268**	**3805**
Quantitative literacy							
Excellent	11	24	37	28	100	293	1834
Good	23	35	29	13	100	263	1487
Moderate	55	25	16	4	100	219	392
Poor	87	8	5	0	100	120	77
No opinion	[13]	-	[2]	-	[15]	*	15
Total	**23**	**28**	**30**	**19**	**100**	**267**	**3805**

Table A4.14 **Literacy level by self-assessment of writing skills for daily life**

	Level 1	Level 2	Level 3	Level 4/5	Total	Mean score	Base
	%	%	%	%	%		
Prose literacy							
Excellent	6	20	43	31	100	301	1356
Good	18	36	33	13	100	270	1655
Moderate	44	38	14	3	100	228	621
Poor	72	22	5	1	100	169	155
No opinion	[15]	[3]	-	-	[18]	*	18
Total	**22**	**30**	**31**	**17**	**100**	**267**	**3805**
Document literacy							
Excellent	8	20	41	32	100	300	1356
Good	20	31	31	18	100	272	1655
Moderate	45	34	16	5	100	230	621
Poor	73	16	10	1	100	165	155
No opinion	[16]	-	[2]	-	[18]	*	18
Total	**23**	**27**	**31**	**19**	**100**	**268**	**3805**
Quantitative literacy							
Excellent	8	21	39	32	100	299	1356
Good	20	32	32	16	100	270	1655
Moderate	43	34	18	6	100	233	621
Poor	73	16	9	2	100	170	155
No opinion	[16]	-	[2]	-	[18]	*	18
Total	**23**	**28**	**30**	**19**	**100**	**267**	**3805**

Table A4.15 **Literacy level by self-assessment of mathematics skills for daily life**

	Level 1	Level 2	Level 3	Level 4/5	Total	Mean score	Base
	%	%	%	%	%		
Prose literacy							
Excellent	8	22	42	28	100	296	1032
Good	15	33	34	18	100	275	1671
Moderate	37	37	20	6	100	238	872
Poor	65	23	10	2	100	193	214
No opinion	[13]	[2]	-	-	[15]	*	15
Total	**22**	**30**	**31**	**17**	**100**	**267**	**3804**
Document literacy							
Excellent	7	16	39	37	100	306	1032
Good	17	31	35	18	100	275	1671
Moderate	40	33	20	6	100	235	872
Poor	70	24	5	2	100	180	214
No opinion	[13]	-	[2]	-	[15]	*	15
Total	**23**	**27**	**31**	**19**	**100**	**268**	**3804**
Quantitative literacy							
Excellent	5	19	38	38	100	309	1032
Good	15	31	37	17	100	278	1671
Moderate	46	34	17	4	100	228	872
Poor	76	19	3	1	100	171	214
No opinion	[13]	-	[2]	-	[15]	*	15
Total	**23**	**28**	**30**	**19**	**100**	**267**	**3804**

Table A4.16 Percentage of respondents who reported various family literacy practices

People who had children aged 6 to 15 living with them

| | Children of respondent ... | | | |
	often see respondent or spouse reading	have time set aside for reading	are limited in amount of time allowed to watch TV	*Base*
	Percentage of respondents			
Prose literacy				
Level 1	68	46	48	*173*
Level 2	89	55	55	*291*
Level 3	96	47	54	*309*
Level 4/5	99	52	46	*158*
Total	**89**	**50**	**51**	*931*
Document literacy				
Level 1	71	46	57	*207*
Level 2	90	55	54	*266*
Level 3	95	43	51	*275*
Level 4/5	100	60	42	*183*
Total	**89**	**50**	**50**	*931*
Quantitative literacy				
Level 1	74	46	56	*204*
Level 2	89	57	51	*274*
Level 3	93	46	54	*276*
Level 4/5	99	51	44	*177*
Total	**89**	**50**	**50**	*931*

Annex tables

A5.1-A5.21

Table A5.1 **Age-group by literacy level and sex**

	16-25		26-35		36-45		46-55		56-65		Total	Base
	%	s.e	%	s.e	%	s.e	%	s.e	%	s.e	%	
Prose literacy												
Men												
Level 1	18		19		17		17		28		100	357
Level 2	21		20		22		16		20		100	510
Level 3	21		25		24		19		10		100	569
Level 4/5	21		26		29		19		5		100	294
Total	**20**		**23**		**23**		**18**		**16**		**100**	**1730**
Women												
Level 1	14		18		18		21		28		100	468
Level 2	19		24		20		19		18		100	655
Level 3	21		25		23		22		9		100	635
Level 4/5	27		28		28		12		6		100	323
Total	**19**		**24**		**22**		**20**		**16**		**100**	**2081**
All												
Level 1	16	1.4	19	1.6	18	1.5	19	1.5	28	1.6	100	825
Level 2	20	1.5	22	1.5	21	1.7	18	1.9	19	1.4	100	1165
Level 3	21	1.6	25	1.6	23	1.2	21	2.4	10	1.0	100	1204
Level 4/5	24	1.9	27	2.2	28	1.8	16	1.5	6	1.1	100	617
Total	**20**	**0.6**	**23**	**0.9**	**22**	**0.8**	**19**	**1.3**	**16**	**0.6**	**100**	**3811**
Document literacy												
Men												
Level 1	16		21		19		17		28		100	343
Level 2	20		19		21		17		23		100	437
Level 3	23		24		23		18		12		100	548
Level 4/5	21		27		28		19		5		100	402
Total	**20**		**23**		**23**		**18**		**16**		**100**	**1730**
Women												
Level 1	15		18		18		21		28		100	557
Level 2	19		24		19		22		16		100	620
Level 3	21		25		23		20		10		100	606
Level 4/5	24		30		29		11		5		100	298
Total	**19**		**24**		**22**		**20**		**16**		**100**	**2081**
All												
Level 1	15	1.4	19	1.7	18	1.6	20	1.5	28	1.5	100	900
Level 2	20	1.2	22	2.1	20	2	19	1.6	19	1.3	100	1057
Level 3	22	1.8	24	1.5	23	1.4	19	2.6	11	0.9	100	1154
Level 4/5	22	1.8	28	1.6	28	1.5	16	1.3	5	0.8	100	700
Total	**20**	**0.6**	**23**	**0.9**	**22**	**0.8**	**19**	**1.3**	**16**	**0.6**	**100**	**3811**
Quantitative literacy												
Men												
Level 1	20		20		18		16		26		100	314
Level 2	24		22		19		14		21		100	428
Level 3	21		20		25		20		13		100	546
Level 4/5	16		29		29		19		7		100	442
Total	**20**		**23**		**23**		**18**		**16**		**100**	**1730**
Women												
Level 1	19		19		18		21		23		100	591
Level 2	18		24		20		21		17		100	642
Level 3	21		26		22		19		11		100	600
Level 4/5	20		28		32		14		6		100	248
Total	**19**		**24**		**22**		**20**		**16**		**100**	**2081**
All												
Level 1	19	1.4	20	1.7	18	1.3	19	1.5	24	1.7	100	905
Level 2	21	1.7	23	1.6	19	1.9	17	1.4	19	1.3	100	1070
Level 3	21	1.7	23	1.6	24	1.3	20	2.5	12	1.1	100	1146
Level 4/5	17	1.9	28	1.8	30	1.4	18	1.7	7	0.8	100	690
Total	**20**	**0.6**	**23**	**0.9**	**22**	**0.8**	**19**	**1.3**	**16**	**0.6**	**100**	**3811**

s.e. = Standard Error of the estimate. The reported sample estimate can be said to be within 2 standard errors of the true population value with 95% confidence.
The standard errors were not available for men and women separately.

Table A5.2 **Highest level of educational attainment by literacy level by sex**

	Primary education or lower		Second level, 1st stage		Second level, 2nd stage		Third level, non-university		Third level, university		Total	Base
	%	s.e	%	s.e	%	s.e	%	s.e	%	s.e	%	
Prose literacy												
Men												
Level 1	17	2.2	67	3.0	14	2.2	1	0.3	2	0.7	100	*357*
Level 2	7	1.2	53	1.9	28	1.6	6	1.0	6	1.3	100	*510*
Level 3	2	0.9	39	2.4	32	1.8	11	1.3	17	1.6	100	*569*
Level 4/5	1	0.9	26	2.8	25	3.1	11	1.3	37	2.9	100	*294*
Total	**6**	**0.5**	**47**	**0.5**	**26**	**0.1**	**7**	**0.5**	**14**	**0.5**	**100**	*1730*
Women												
Level 1	14	2.1	74	3.2	8	3.0	2	0.8	2	0.7	100	*468*
Level 2	4	1.1	76	4.5	12	4.8	5	1.1	3	0.6	100	*655*
Level 3	4	1.2	56	4.6	15	6.0	14	1.5	11	1.5	100	*635*
Level 4/5	3	1.4	34	3.6	25	5.3	13	2.1	25	2.4	100	*323*
Total	**6**	**0.8**	**62**	**3.6**	**14**	**4.7**	**9**	**0.6**	**9**	**0.6**	**100**	*2081*
All												
Level 1	15	1.5	70	2.2	11	1.9	2	0.5	2	0.5	100	*825*
Level 2	5	0.8	65	2.3	20	2.4	6	0.7	5	0.7	100	*1165*
Level 3	3	0.7	47	2.2	23	2.7	12	0.9	14	1.1	100	*1204*
Level 4/5	2	0.9	29	2.2	25	2.8	12	1.3	31	1.8	100	*617*
Total	**6**	**0.5**	**55**	**1.6**	**20**	**2.1**	**8**	**0.4**	**11**	**0.4**	**100**	*3811*
Document literacy												
Men												
Level 1	16	2.2	66	3.1	15	1.9	1	0.4	2	0.8	100	*343*
Level 2	7	1.3	55	1.6	28	2.1	6	1.3	5	1.1	100	*437*
Level 3	4	1.2	43	2.1	29	1.9	10	1.5	15	2.0	100	*548*
Level 4/5	1	0.7	27	2.4	29	2.3	11	1.1	32	2.6	100	*402*
Total	**6**	**0.5**	**47**	**0.5**	**26**	**0.1**	**7**	**0.5**	**14**	**0.5**	**100**	*1730*
Women												
Level 1	12	1.8	74	3.5	9	3.5	4	1.0	1	0.6	100	*557*
Level 2	5	1.3	72	3.7	11	4.0	6	1.0	5	0.9	100	*620*
Level 3	3	1.1	56	5.0	17	6.8	14	1.6	11	1.4	100	*606*
Level 4/5	4	1.7	36	3.7	24	4.8	12	2.1	24	2.5	100	*298*
Total	**6**	**0.8**	**62**	**3.6**	**14**	**4.7**	**9**	**0.6**	**9**	**0.6**	**100**	*2081*
All												
Level 1	14	1.4	71	2.3	11	2.1	2	0.6	2	0.5	100	*900*
Level 2	6	0.8	64	1.9	19	2.1	6	0.7	5	0.7	100	*1057*
Level 3	3	0.8	49	2.4	23	3.2	12	1.0	13	1.1	100	*1154*
Level 4/5	2	0.8	30	2.1	27	2.3	11	1.0	29	2.0	100	*700*
Total	**6**	**0.5**	**55**	**1.6**	**20**	**2.1**	**8**	**0.4**	**11**	**0.4**	**100**	*3811*
Quantitative literacy												
Men												
Level 1	18	2.4	65	3.3	15	2.2	1	0.5	2	0.6	100	*314*
Level 2	8	1.4	54	2.3	28	2.1	5	1.1	5	1.1	100	*428*
Level 3	4	1.2	46	2.1	31	1.7	9	1.2	11	1.7	100	*546*
Level 4/5	1	0.6	27	2.0	25	2.0	12	1.1	35	2.2	100	*442*
Total	**6**	**0.5**	**47**	**0.5**	**26**	**0.1**	**7**	**0.5**	**14**	**0.5**	**100**	*1730*
Women												
Level 1	11	1.4	76	3.4	9	3.6	3	0.9	1	0.5	100	*591*
Level 2	7	1.3	69	3.1	11	3.3	7	1.0	5	0.9	100	*642*
Level 3	2	0.8	58	5.2	17	6.6	13	1.4	11	1.6	100	*600*
Level 4/5	2	1.4	26	3.9	27	7.2	14	2.7	31	3.9	100	*248*
Total	**6**	**0.8**	**62**	**3.6**	**14**	**4.7**	**9**	**0.6**	**9**	**0.6**	**100**	*2081*
All												
Level 1	13	1.2	71	2.2	11	2.2	2	0.6	1	0.4	100	*905*
Level 2	7	0.8	62	1.9	19	1.9	6	0.7	5	0.7	100	*1070*
Level 3	3	0.7	52	2.6	24	3.1	11	0.8	11	1.1	100	*1146*
Level 4/5	1	0.7	26	1.8	26	2.7	12	1.1	34	2.0	100	*690*
Total	**6**	**0.5**	**55**	**1.6**	**20**	**2.1**	**8**	**0.4**	**11**	**0.4**	**100**	*3811*

s.e = Standard Error of the estimate. The reported sample estimate can be said to be within 2 standard errors of the true population value with 95% confidence.

Table A5.3 **Economic activity status by literacy level**

	Employed	Unemployed	Student	Home duties	Retired	Other inactive	Total	Base
	%	%	%	%	%	%	%	
Prose literacy								
Level 1	51	14	1	11	8	15	100	825
Level 2	69	8	3	8	5	7	100	1165
Level 3	78	8	4	5	3	2	100	1204
Level 4/5	82	7	5	3	2	1	100	617
Total	**70**	**9**	**3**	**7**	**5**	**6**	**100**	**3811**
Document literacy								
Level 1	50	13	1	12	9	16	100	900
Level 2	67	11	3	8	5	6	100	1057
Level 3	78	7	4	5	3	2	100	1154
Level 4/5	84	6	4	2	2	1	100	700
Total	**70**	**9**	**3**	**7**	**5**	**6**	**100**	**3811**
Quantitative literacy								
Level 1	50	13	1	13	8	15	100	905
Level 2	68	10	4	7	5	6	100	1070
Level 3	76	8	3	5	4	3	100	1146
Level 4/5	86	4	4	2	2	1	100	690
Total	**70**	**9**	**3**	**7**	**5**	**6**	**100**	**3811**

Table A5.4 **Social class by literacy level**

	I and II	III (non-manual)	III (manual)	IV and V	Total	Bases*
	%	%	%	%	%	
Prose literacy						
Level 1	14	13	26	46	100	731
Level 2	22	24	24	30	100	1089
Level 3	41	27	16	16	100	1150
Level 4/5	51	29	6	13	100	599
Total	**31**	**24**	**19**	**26**	**100**	**3569**
Document literacy						
Level 1	14	16	26	44	100	806
Level 2	23	24	24	29	100	979
Level 3	38	27	15	20	100	1108
Level 4/5	52	27	10	11	100	676
Total	**31**	**24**	**19**	**26**	**100**	**3569**
Quantitative literacy						
Level 1	13	18	23	46	100	808
Level 2	21	24	25	30	100	986
Level 3	38	28	15	18	100	1105
Level 4/5	57	22	11	10	100	670
Total	**31**	**24**	**19**	**26**	**100**	**3569**

* Social class was not derived for Members of the Armed Forces, persons in inadequately described occupations and persons who had never worked.

Table A5.5 **Whether or not receives social security benefits (excluding pensions and child benefit) by literacy level**

	Yes	No	Total	Base
	%	%	%	
Prose literacy				
Level 1	37	63	100	812
Level 2	21	79	100	1159
Level 3	14	86	100	1197
Level 4/5	9	91	100	616
Total	20	80	100	3784
Document literacy				
Level 1	38	62	100	886
Level 2	20	80	100	1051
Level 3	13	87	100	1148
Level 4/5	8	92	100	699
Total	20	80	100	3784
Quantitative literacy				
Level 1	39	61	100	893
Level 2	19	81	100	1062
Level 3	15	85	100	1140
Level 4/5	6	94	100	689
Total	20	80	100	3784

Table A5.6 **Amount of time spent watching televsion by literacy level**

	Not on a daily basis	1 hour or less per day	1-2 hours per day	More than 2 hours but less than 5	5 or more hours per day	Do not have a television	Total	Base
	%	%	%	%	%	%	%	
Prose literacy								
Level 1	5	5	19	45	25	0	100	821
Level 2	3	7	23	51	15	1	100	1161
Level 3	4	10	29	50	6	1	100	1204
Level 4/5	5	17	34	40	4	0	100	617
Total	4	9	26	47	13	1	100	3803
Document literacy								
Level 1	4	7	19	45	25	0	100	896
Level 2	4	6	22	51	16	1	100	1054
Level 3	4	11	28	51	6	1	100	1153
Level 4/5	5	15	36	40	5	0	100	700
Total	4	9	26	47	13	1	100	3803
Quantitative literacy								
Level 1	4	6	18	44	26	0	100	901
Level 2	4	7	22	51	14	1	100	1066
Level 3	4	10	28	50	7	1	100	1146
Level 4/5	4	16	37	40	3	0	100	690
Total	4	9	26	47	13	1	100	3803

Table A5.7 **Frequency of reading books by literacy level and sex**

	Daily	Weekly	Monthly	Several times a year	Never	Total	Base
	%	%	%	%	%	%	
Prose literacy							
Men							
Level 1	9	10	9	21	51	100	356
Level 2	22	14	16	26	22	100	507
Level 3	32	18	13	26	12	100	569
Level 4/5	39	24	9	22	5	100	294
Total	**25**	**16**	**13**	**24**	**22**	**100**	1726
Women							
Level 1	22	11	14	23	30	100	466
Level 2	34	17	13	22	14	100	652
Level 3	44	16	14	19	7	100	635
Level 4/5	54	21	10	11	4	100	323
Total	**37**	**16**	**13**	**19**	**14**	**100**	2076
All							
Level 1	16	10	12	22	40	100	822
Level 2	28	16	15	24	18	100	1159
Level 3	38	17	14	23	9	100	1204
Level 4/5	46	23	10	17	4	100	617
Total	**31**	**16**	**13**	**22**	**18**	**100**	3802
Document literacy							
Men							
Level 1	9	12	11	17	51	100	342
Level 2	19	13	16	28	24	100	434
Level 3	31	17	12	25	14	100	548
Level 4/5	37	22	11	25	6	100	402
Total	**25**	**16**	**13**	**24**	**22**	**100**	1726
Women							
Level 1	22	12	14	24	28	100	555
Level 2	37	17	16	19	11	100	618
Level 3	44	16	12	19	9	100	605
Level 4/5	53	20	9	13	5	100	298
Total	**37**	**16**	**13**	**19**	**14**	**100**	2076
All							
Level 1	17	12	13	21	38	100	897
Level 2	29	15	16	23	17	100	1052
Level 3	37	17	12	22	11	100	1153
Level 4/5	43	21	10	20	5	100	700
Total	**31**	**16**	**13**	**22**	**18**	**100**	3802
Quantitative literacy							
Men							
Level 1	11	11	12	15	50	100	313
Level 2	21	12	15	28	24	100	425
Level 3	29	19	12	24	16	100	546
Level 4/5	36	20	11	26	7	100	442
Total	**25**	**16**	**13**	**24**	**22**	**100**	1726
Women							
Level 1	23	14	14	24	24	100	589
Level 2	37	14	17	20	12	100	639
Level 3	48	17	10	16	9	100	600
Level 4/5	49	22	11	14	4	100	248
Total	**37**	**16**	**13**	**19**	**14**	**100**	2076
All							
Level 1	18	13	14	21	34	100	902
Level 2	29	13	16	24	18	100	1064
Level 3	38	18	11	20	12	100	1146
Level 4/5	40	21	11	23	6	100	690
Total	**31**	**16**	**13**	**22**	**18**	**100**	3802

Table A5.8 **Frequency of reading newspapers or magazines by literacy level**

	Daily	Weekly	Monthly	Several times a year	Never	Total	Base
	%	%	%	%	%	%	
Prose literacy							
Level 1	69	19	3	2	7	100	824
Level 2	75	19	3	2	1	100	1159
Level 3	74	21	3	1	1	100	1204
Level 4/5	67	31	2	1	1	100	617
Total	72	22	3	1	2	100	3804
Document literacy							
Level 1	70	18	3	2	7	100	899
Level 2	75	20	4	1	1	100	1052
Level 3	72	22	3	1	1	100	1153
Level 4/5	69	28	1	1	0	100	700
Total	72	22	3	1	2	100	3804
Quantitative literacy							
Level 1	68	20	3	2	7	100	904
Level 2	76	19	3	1	1	100	1064
Level 3	71	23	3	1	1	100	1146
Level 4/5	71	26	2	1	0	100	690
Total	72	22	3	1	2	100	3804

Table A 5.9 **Sections of the newspaper read by literacy level**

	Do not read newspaper	Classified ads	Other ads	Nat/int news	Regional/local news	Sports	Home, fashion, health	Editorial	Financial/shares	Comics	TV listings	Film, concert listings	Book, film or art reviews	Horoscope/stars	Advice column	Special interest sections	Personal finance	Other	Base
	%	%	%	%	%	%	%	%	%	%	%	%	%	%	%	%	%	%	
Prose literacy																			
Level 1	8	43	28	54	65	39	36	15	15	17	65	25	17	46	26	23	17	3	824
Level 2	1	47	36	72	79	46	49	28	26	22	74	41	32	44	36	36	30	2	1159
Level 3	1	42	33	86	83	52	51	39	33	21	77	44	44	37	31	51	37	1	1204
Level 4/5	1	33	34	92	85	54	56	44	35	17	81	49	56	31	31	57	41	2	617
Total	3	42	33	76	78	48	48	31	27	19	74	40	37	40	32	42	31	2	3804
Document literacy																			
Level 1	7	41	28	57	66	38	40	17	14	16	67	26	18	47	29	21	17	3	899
Level 2	2	49	36	73	81	45	51	30	25	21	74	43	36	46	36	37	29	3	1052
Level 3	1	42	34	84	82	49	51	37	33	20	77	43	40	37	31	50	36	1	1153
Level 4/5	1	35	34	90	84	61	48	42	39	20	79	47	53	29	29	61	45	2	700
Total	3	42	33	76	78	48	48	31	27	19	74	40	37	40	32	42	31	2	3804
Quantitative literacy																			
Level 1	7	42	29	53	66	34	43	16	13	16	67	29	21	50	31	21	16	3	904
Level 2	1	46	34	75	81	46	50	28	24	20	76	42	36	45	36	38	27	2	1064
Level 3	2	44	34	84	83	48	53	38	31	20	74	42	40	38	32	49	36	2	1146
Level 4/5	0	35	33	93	84	67	44	45	46	22	80	46	51	24	26	61	51	1	690
Total	3	42	33	76	78	48	48	31	27	19	74	40	37	40	32	42	31	2	3804

Percentages add to more than 100% because respondents were asked to list all the sections of a newspaper they generally read.

Table A5.10 **How often respondent follows current events by literacy level and sex**

	Most of the time	Some of the time	Only now and then	Hardly at all	Total	Base
	%	%	%	%	%	
Prose literacy						
Men						
Level 1	36	27	21	17	100	356
Level 2	52	27	15	6	100	509
Level 3	60	31	8	1	100	569
Level 4/5	69	24	6	0	100	294
Total	54	28	12	6	100	1728
Women						
Level 1	25	28	28	19	100	464
Level 2	36	36	18	10	100	652
Level 3	48	36	12	5	100	635
Level 4/5	53	37	6	4	100	323
Total	40	34	16	9	100	2074
All						
Level 1	30	27	24	18	100	820
Level 2	44	32	16	8	100	1161
Level 3	54	33	10	3	100	1204
Level 4/5	61	30	6	2	100	617
Total	47	31	14	8	100	3802
Document literacy						
Men						
Level 1	38	26	19	17	100	342
Level 2	51	27	17	6	100	436
Level 3	57	31	9	3	100	548
Level 4/5	67	27	6	0	100	402
Total	54	28	12	6	100	1728
Women						
Level 1	27	29	25	18	100	553
Level 2	41	34	17	9	100	618
Level 3	43	40	11	6	100	605
Level 4/5	53	35	9	2	100	298
Total	40	34	16	9	100	2074
All						
Level 1	32	28	22	18	100	895
Level 2	46	30	17	7	100	1054
Level 3	50	35	10	4	100	1153
Level 4/5	62	30	7	1	100	700
Total	47	31	14	8	100	3802
Quantitative literacy						
Men						
Level 1	35	26	21	19	100	313
Level 2	49	28	18	5	100	427
Level 3	57	31	8	3	100	546
Level 4/5	69	26	5	1	100	442
Total	54	28	12	6	100	1728
Women						
Level 1	26	29	26	19	100	587
Level 2	39	35	17	8	100	639
Level 3	47	38	10	5	100	600
Level 4/5	56	35	7	2	100	248
Total	40	34	16	9	100	2074
All						
Level 1	30	28	24	19	100	900
Level 2	44	32	18	7	100	1066
Level 3	52	35	9	4	100	1146
Level 4/5	65	29	6	1	100	690
Total	47	31	14	8	100	3802

Table A5.11 **Satisfaction with reading and writing skills in daily life by literacy level and sex**

	Very satisfied	Somewhat satisfied	Somewhat dissatisfied	Very dissatisfied	No opinion	Total	Base
	%	%	%	%	%	%	
Prose literacy							
Men							
Level 1	27	43	15	10	4	100	356
Level 2	37	49	12	1	0	100	509
Level 3	55	40	5	0	-	100	569
Level 4/5	71	26	3	-	-	100	294
Total	47	41	9	3	1	100	1728
Women							
Level 1	29	46	16	5	4	100	464
Level 2	45	45	9	1	0	100	652
Level 3	67	30	3	-	0	100	635
Level 4/5	83	17	1	-	-	100	323
Total	54	36	7	2	1	100	2074
All							
Level 1	28	45	16	8	4	100	820
Level 2	41	47	10	1	0	100	1161
Level 3	61	35	4	0	0	100	1204
Level 4/5	77	22	2	0	0	100	617
Total	50	39	8	2	1	100	3802
Document literacy							
Men							
Level 1	26	43	16	11	4	100	342
Level 2	40	48	11	1	-	100	436
Level 3	51	41	7	0	0	100	548
Level 4/5	65	32	3	-	-	100	402
Total	47	41	9	3	1	100	1728
Women							
Level 1	32	46	14	5	3	100	553
Level 2	51	40	8	1	0	100	618
Level 3	65	31	3	-	0	100	605
Level 4/5	78	21	1	-	-	100	298
Total	54	36	7	2	1	100	2074
All							
Level 1	29	45	15	7	4	100	895
Level 2	46	44	10	1	0	100	1054
Level 3	58	37	5	0	0	100	1153
Level 4/5	70	28	2	-	-	100	700
Total	50	39	8	2	1	100	3802
Quantitative literacy							
Men							
Level 1	25	41	16	12	5	100	313
Level 2	40	48	11	1	-	100	427
Level 3	49	43	7	1	0	100	546
Level 4/5	65	31	4	-	-	100	442
Total	47	41	9	3	1	100	1728
Women							
Level 1	32	46	15	4	3	100	587
Level 2	51	42	6	1	0	100	639
Level 3	68	30	2	-	0	100	600
Level 4/5	81	16	2	-	-	100	248
Total	54	36	7	2	1	100	2074
All							
Level 1	29	44	16	7	4	100	900
Level 2	46	45	8	1	0	100	1066
Level 3	58	36	5	0	0	100	1146
Level 4/5	70	26	3	-	-	100	690
Total	50	39	8	2	1	100	3802

Table A5.12 **Self-assessment of reading skills in daily life by literacy level**

	Excellent	Good	Moderate	Poor	No opinion	Total	Base
	%	%	%	%	%	%	
Prose literacy							
Men							
Level 1	16	40	30	11	2	100	356
Level 2	36	49	14	1	0	100	509
Level 3	57	38	4	1	-	100	569
Level 4/5	70	29	1	-	-	100	294
Total	**45**	**40**	**12**	**3**	**1**	**100**	1728
Women							
Level 1	20	44	25	9	2	100	467
Level 2	39	50	10	1	0	100	652
Level 3	60	37	3	0	-	100	635
Level 4/5	76	23	1	-	-	100	323
Total	**47**	**40**	**10**	**2**	**1**	**100**	2077
All							
Level 1	18	42	28	10	2	100	823
Level 2	37	50	12	1	0	100	1161
Level 3	59	38	3	1	-	100	1204
Level 4/5	73	26	1	-	-	100	617
Total	**46**	**40**	**11**	**3**	**1**	**100**	3805
Document literacy							
Men							
Level 1	15	40	31	12	2	100	342
Level 2	35	51	14	0	-	100	436
Level 3	57	36	6	1	0	100	548
Level 4/5	64	34	2	-	-	100	402
Total	**45**	**40**	**12**	**3**	**1**	**100**	1728
Women							
Level 1	24	43	23	8	2	100	556
Level 2	45	46	9	0	-	100	618
Level 3	59	39	3	0	-	100	605
Level 4/5	71	28	2	-	-	100	298
Total	**47**	**40**	**10**	**2**	**1**	**100**	2077
All							
Level 1	20	42	26	10	2	100	898
Level 2	40	48	11	0	-	100	1054
Level 3	58	37	4	1	0	100	1153
Level 4/5	67	31	2	-	-	100	700
Total	**46**	**40**	**11**	**3**	**1**	**100**	3805
Quantitative literacy							
Men							
Level 1	15	37	31	13	3	100	313
Level 2	36	51	12	1	-	100	427
Level 3	50	40	9	1	0	100	546
Level 4/5	68	30	2	0	-	100	442
Total	**45**	**40**	**12**	**3**	**1**	**100**	1728
Women							
Level 1	26	43	22	7	2	100	590
Level 2	42	49	8	1	-	100	639
Level 3	62	36	2	-	-	100	600
Level 4/5	72	26	2	-	-	100	248
Total	**47**	**40**	**10**	**2**	**1**	**100**	2077
All							
Level 1	22	41	26	10	2	100	903
Level 2	39	50	10	1	-	100	1066
Level 3	56	38	6	0	0	100	1146
Level 4/5	69	29	2	0	-	100	690
Total	**46**	**40**	**11**	**3**	**1**	**100**	3805

Table A5.13 **Self-assessment of writing skills in daily life by literacy level**

	Excellent	Good	Moderate	Poor	No opinion	Total	Base
	%	%	%	%	%	%	
Prose literacy							
Men							
Level 1	7	30	38	22	3	100	356
Level 2	19	49	26	5	0	100	509
Level 3	43	47	9	1	-	100	569
Level 4/5	57	38	5	1	-	100	294
Total	**31**	**43**	**20**	**7**	**1**	**100**	*1728*
Women							
Level 1	11	44	31	12	2	100	467
Level 2	25	57	16	2	0	100	652
Level 3	46	48	6	0	-	100	635
Level 4/5	65	34	2	-	-	100	323
Total	**35**	**47**	**14**	**3**	**1**	**100**	*2077*
All							
Level 1	9	37	34	17	3	100	823
Level 2	22	53	21	4	0	100	1161
Level 3	44	47	8	1	-	100	1204
Level 4/5	61	36	3	0	-	100	617
Total	**33**	**45**	**17**	**5**	**1**	**100**	*3805*
Document literacy							
Men							
Level 1	7	28	39	23	3	100	342
Level 2	18	50	28	5	-	100	436
Level 3	41	46	11	3	0	100	548
Level 4/5	50	43	6	0	-	100	402
Total	**31**	**43**	**20**	**7**	**1**	**100**	*1728*
Women							
Level 1	14	45	28	11	2	100	556
Level 2	29	54	15	2	-	100	618
Level 3	47	47	6	0	-	100	605
Level 4/5	58	40	2	-	-	100	298
Total	**35**	**47**	**14**	**3**	**1**	**100**	*2077*
All							
Level 1	11	38	32	16	3	100	898
Level 2	24	52	21	3	-	100	1054
Level 3	43	46	9	2	0	100	1153
Level 4/5	54	42	5	0	-	100	700
Total	**33**	**45**	**17**	**5**	**1**	**100**	*3805*
Quantitative literacy							
Men							
Level 1	7	28	38	24	3	100	313
Level 2	22	49	25	5	-	100	427
Level 3	33	49	14	3	0	100	546
Level 4/5	53	39	7	1	-	100	442
Total	**31**	**43**	**20**	**7**	**1**	**100**	*1728*
Women							
Level 1	15	47	26	10	2	100	590
Level 2	28	54	16	1	-	100	639
Level 3	50	45	5	0	-	100	600
Level 4/5	60	39	1	-	-	100	248
Total	**35**	**47**	**14**	**3**	**1**	**100**	*2077*
All							
Level 1	12	39	31	16	3	100	903
Level 2	25	52	21	3	-	100	1066
Level 3	42	47	10	2	0	100	1146
Level 4/5	55	39	5	1	-	100	690
Total	**33**	**45**	**17**	**5**	**1**	**100**	*3805*

Table A5.14 **Self-assessment of mathematics skills in daily life by literacy level**

	Excellent	Good	Moderate	Poor	No opinion	Total	Base
	%	%	%	%	%	%	
Prose literacy							
Men							
Level 1	12	34	38	14	2	100	356
Level 2	21	53	23	3	0	100	509
Level 3	42	47	9	1	-	100	569
Level 4/5	49	44	6	1	-	100	294
Total	**31**	**45**	**19**	**4**	**1**	**100**	1728
Women							
Level 1	6	31	42	20	2	100	467
Level 2	16	45	34	5	-	100	652
Level 3	26	52	20	2	-	100	635
Level 4/5	36	53	10	1	-	100	323
Total	**20**	**45**	**28**	**7**	**0**	**100**	2077
All							
Level 1	9	32	40	17	2	100	823
Level 2	19	49	28	4	0	100	1161
Level 3	34	49	15	2	-	100	1204
Level 4/5	43	49	8	1	-	100	617
Total	**25**	**45**	**23**	**6**	**0**	**100**	3805
Document literacy							
Men							
Level 1	10	34	39	15	2	100	342
Level 2	17	55	24	4	-	100	436
Level 3	38	49	12	0	0	100	548
Level 4/5	54	40	5	1	-	100	402
Total	**31**	**45**	**19**	**4**	**1**	**100**	1728
Women							
Level 1	7	32	41	18	1	100	556
Level 2	14	48	32	5	-	100	618
Level 3	26	54	18	1	-	100	605
Level 4/5	42	45	12	0	-	100	298
Total	**20**	**45**	**28**	**7**	**0**	**100**	2077
All							
Level 1	8	33	40	17	2	100	898
Level 2	15	51	29	5	-	100	1054
Level 3	33	51	15	1	0	100	1153
Level 4/5	49	42	8	1	-	100	700
Total	**25**	**45**	**23**	**6**	**0**	**100**	3805
Quantitative literacy							
Men							
Level 1	6	29	45	17	3	100	313
Level 2	19	52	26	4	-	100	427
Level 3	35	54	10	1	0	100	546
Level 4/5	56	40	4	0	-	100	442
Total	**31**	**45**	**19**	**4**	**1**	**100**	1728
Women							
Level 1	5	29	46	19	1	100	590
Level 2	15	49	31	4	-	100	639
Level 3	29	55	15	0	-	100	600
Level 4/5	46	48	6	0	-	100	248
Total	**20**	**45**	**28**	**7**	**0**	**100**	2077
All							
Level 1	5	29	45	18	2	100	903
Level 2	17	51	29	4	-	100	1066
Level 3	32	55	13	1	0	100	1146
Level 4/5	53	42	4	0	-	100	690
Total	**25**	**45**	**23**	**6**	**0**	**100**	3805

Table A5.15 How often help is required filling out forms by literacy level

	Often	Sometimes	Never	Total	Base
	%	%	%	%	
Prose literacy					
Men					
Level 1	19	30	51	100	356
Level 2	2	20	78	100	508
Level 3	1	10	89	100	569
Level 4/5	-	8	92	100	294
Total	5	17	78	100	1727
Women					
Level 1	14	29	57	100	466
Level 2	3	19	78	100	652
Level 3	1	12	87	100	635
Level 4/5	1	6	93	100	323
Total	4	17	79	100	2076
All					
Level 1	16	29	54	100	822
Level 2	2	20	78	100	1160
Level 3	1	11	88	100	1204
Level 4/5	0	7	93	100	617
Total	5	17	79	100	3803
Document literacy					
Men					
Level 1	18	34	48	100	342
Level 2	3	18	79	100	435
Level 3	2	11	87	100	548
Level 4/5	-	9	91	100	402
Total	5	17	78	100	1727
Women					
Level 1	13	29	58	100	555
Level 2	2	18	80	100	618
Level 3	1	10	89	100	605
Level 4/5	0	7	93	100	298
Total	4	17	79	100	2076
All					
Level 1	15	31	54	100	897
Level 2	2	18	80	100	1053
Level 3	1	10	88	100	1153
Level 4/5	0	8	92	100	700
Total	5	17	79	100	3803
Quantitative literacy					
Men					
Level 1	20	34	46	100	313
Level 2	4	20	76	100	427
Level 3	1	13	87	100	545
Level 4/5	-	7	93	100	442
Total	5	17	78	100	1727
Women					
Level 1	12	29	60	100	589
Level 2	3	19	78	100	639
Level 3	0	8	92	100	600
Level 4/5	0	5	94	100	248
Total	4	17	79	100	2076
All					
Level 1	15	31	54	100	902
Level 2	3	20	77	100	1066
Level 3	1	10	89	100	1145
Level 4/5	0	6	94	100	690
Total	5	17	79	100	3803

Table A5.16 How often help is required reading information from government departments, businesses or other institutions by literacy level

	Often	Sometimes	Never	Total	Base
	%	%	%	%	
Prose literacy					
Men					
Level 1	17	24	58	100	355
Level 2	1	25	74	100	509
Level 3	1	19	80	100	569
Level 4/5	1	10	89	100	294
Total	4	20	75	100	1727
Women					
Level 1	11	28	61	100	466
Level 2	2	26	72	100	651
Level 3	2	23	75	100	635
Level 4/5	1	18	81	100	323
Total	4	24	72	100	2075
All					
Level 1	14	26	60	100	821
Level 2	1	25	73	100	1160
Level 3	2	21	77	100	1204
Level 4/5	1	14	85	100	617
Total	4	22	74	100	3802
Document literacy					
Men					
Level 1	19	25	56	100	341
Level 2	1	26	73	100	436
Level 3	1	19	80	100	548
Level 4/5	1	13	87	100	402
Total	4	20	75	100	1727
Women					
Level 1	10	29	61	100	555
Level 2	2	26	73	100	617
Level 3	2	21	77	100	605
Level 4/5	0	19	81	100	298
Total	4	24	72	100	2075
All					
Level 1	14	27	59	100	896
Level 2	1	26	73	100	1053
Level 3	1	20	79	100	1153
Level 4/5	1	15	85	100	700
Total	4	22	74	100	3802
Quantitative literacy					
Men					
Level 1	19	25	56	100	312
Level 2	3	26	71	100	427
Level 3	1	20	79	100	546
Level 4/5	1	11	88	100	442
Total	4	20	75	100	1727
Women					
Level 1	10	30	61	100	589
Level 2	3	27	70	100	638
Level 3	1	18	81	100	600
Level 4/5	0	19	81	100	248
Total	4	24	72	100	2075
All					
Level 1	13	28	59	100	901
Level 2	3	27	71	100	1065
Level 3	1	19	80	100	1146
Level 4/5	1	13	86	100	690
Total	4	22	74	100	3802

Table A5.17 **How often help is required doing basic mathematics by literacy level**

	Often	Sometimes	Never	Total	Base
	%	%	%	%	
Prose literacy					
Men					
Level 1	6	17	77	100	356
Level 2	0	6	93	100	509
Level 3	0	1	99	100	569
Level 4/5	-	1	99	100	294
Total	1	6	93	100	1728
Women					
Level 1	11	18	71	100	466
Level 2	1	9	90	100	652
Level 3	1	3	96	100	635
Level 4/5	-	1	99	100	323
Total	3	8	89	100	2076
All					
Level 1	9	17	74	100	822
Level 2	1	7	92	100	1161
Level 3	0	2	98	100	1204
Level 4/5	-	1	99	100	617
Total	2	7	91	100	3804
Document literacy					
Men					
Level 1	7	18	75	100	342
Level 2	1	7	92	100	436
Level 3	0	1	99	100	548
Level 4/5	-	1	99	100	402
Total	1	6	93	100	1728
Women					
Level 1	10	17	74	100	555
Level 2	1	9	90	100	618
Level 3	0	2	98	100	605
Level 4/5	-	2	98	100	298
Total	3	8	89	100	2076
All					
Level 1	8	17	75	100	897
Level 2	1	8	91	100	1054
Level 3	0	2	98	100	1153
Level 4/5	-	1	99	100	700
Total	2	7	91	100	3804
Quantitative literacy					
Men					
Level 1	8	22	70	100	313
Level 2	0	6	93	100	427
Level 3	-	0	100	100	546
Level 4/5	-	1	99	100	442
Total	1	6	93	100	1728
Women					
Level 1	10	19	71	100	589
Level 2	1	7	93	100	639
Level 3	-	1	99	100	600
Level 4/5	-	0	100	100	248
Total	3	8	89	100	2076
All					
Level 1	9	20	71	100	902
Level 2	1	6	93	100	1066
Level 3	-	1	99	100	1146
Level 4/5	-	1	99	100	690
Total	2	7	91	100	3804

Table A5.18 **How often help is required writing notes and letters by literacy level**

	Often	Sometimes	Never	Total	Base
	%	%	%	%	
Prose literacy					
Level 1	13	17	70	100	823
Level 2	2	10	88	100	1161
Level 3	1	4	95	100	1204
Level 4/5	-	3	97	100	617
Total	4	9	88	100	3805
Document literacy					
Level 1	12	17	72	100	898
Level 2	2	9	89	100	1054
Level 3	1	5	94	100	1153
Level 4/5	-	3	97	100	700
Total	4	9	88	100	3805
Quantitative literacy					
Level 1	12	16	72	100	903
Level 2	2	10	88	100	1066
Level 3	1	5	94	100	1146
Level 4/5	-	4	96	100	690
Total	4	9	88	100	3805

Table A5.19 **How often help is required reading newspapers by literacy level**

	Often	Sometimes	Never	Total	Base
	%	%	%	%	
Prose literacy					
Level 1	7	11	82	100	821
Level 2	0	2	98	100	1161
Level 3	0	1	98	100	1204
Level 4/5	-	0	100	100	617
Total	2	3	95	100	3803
Document literacy					
Level 1	7	9	84	100	896
Level 2	0	3	97	100	1054
Level 3	0	1	99	100	1153
Level 4/5	-	0	100	100	700
Total	2	3	95	100	3803
Quantitative literacy					
Level 1	7	9	84	100	901
Level 2	0	3	97	100	1066
Level 3	0	1	99	100	1146
Level 4/5	0	0	100	100	690
Total	2	3	95	100	3803

Table A5.20 **How often help is required reading instructions on medicine bottles by literacy level**

	Often	Sometimes	Never	Total	Base
	%	%	%	%	
Prose literacy					
Level 1	7	7	87	100	823
Level 2	0	2	98	100	1161
Level 3	-	2	98	100	1204
Level 4/5	-	1	99	100	617
Total	2	3	96	100	3805
Document literacy					
Level 1	7	7	87	100	898
Level 2	0	2	98	100	1054
Level 3	-	2	98	100	1153
Level 4/5	-	2	98	100	700
Total	2	3	96	100	3805
Quantitative literacy					
Level 1	7	7	87	100	903
Level 2	0	3	97	100	1066
Level 3	-	1	99	100	1146
Level 4/5	-	1	99	100	690
Total	2	3	96	100	3805

Table A5.21 **How often help is required reading packaged goods in shops by literacy level**

	Often	Sometimes	Never	Total	Base
	%	%	%	%	
Prose literacy					
Level 1	7	7	87	100	823
Level 2	0	2	98	100	1161
Level 3	-	2	98	100	1204
Level 4/5	-	1	99	100	617
Total	2	3	96	100	3805
Document literacy					
Level 1	6	6	87	100	898
Level 2	0	2	98	100	1054
Level 3	0	2	98	100	1153
Level 4/5	-	2	98	100	700
Total	2	3	96	100	3805
Quantitative literacy					
Level 1	6	7	87	100	903
Level 2	-	3	97	100	1066
Level 3	0	1	98	100	1146
Level 4/5	-	1	99	100	690
Total	2	3	96	100	3805

137

Annex tables
A6.1-A6.6

Annex figures
A6.1a-A6.1d

Table A6.1 **Literacy level by age group and country**

Country		Level 1		Level 2		Level 3		Level 4/5		Total
		%	s.e.	%	s.e.	%	s.e.	%	s.e.	%
Prose literacy										
Canada	16-25	11	1.9	26	3.5	44	2.5	20	4.5	100
	26-35	12	2.5	29	3.7	33	4.9	26	3.3	100
	36-45	13	1.8	19	3.2	37	4.8	31	4.1	100
	46-55	21	4.3	30	2.5	31	4.1	18	5.8	100
	56-65	38	7.8	26	4.0	28	10.1	8	6.0	100
Germany	16-25	9	1.9	29	3.8	46	4.9	15	3.8	100
	26-35	12	1.8	31	1.2	37	1.7	20	2.2	100
	36-45	14	1.8	32	1.8	39	2.6	15	1.5	100
	46-55	14	2.4	37	2.3	38	3.3	11	2.0	100
	56-65	22	2.8	43	3.8	30	2.8	5	1.4	100
Great Britain	16-25	17	1.7	30	2.4	33	2.7	20	1.8	100
	26-35	18	1.5	29	2.0	34	2.0	19	1.5	100
	36-45	17	1.3	29	2.3	33	2.2	21	1.6	100
	46-55	22	2.6	29	2.7	35	2.4	14	1.3	100
	56-65	39	2.2	37	2.2	19	1.5	6	1.2	100
Netherlands	16-25	8	1.8	22	2.4	50	3.4	19	2.2	100
	26-35	6	1.2	21	1.6	51	2.1	23	1.4	100
	36-45	9	1.0	30	1.4	47	1.8	14	1.2	100
	46-55	14	1.6	39	2.1	38	2.4	10	1.5	100
	56-65	20	2.2	48	3.4	28	2.5	5	1.0	100
Poland	16-25	27	2.2	38	2.0	29	1.4	6	1.1	100
	26-35	35	2.4	39	2.3	22	1.5	4	0.8	100
	36-45	42	1.5	38	1.5	17	1.0	3	0.5	100
	46-55	53	1.7	30	1.6	16	2.1	1	0.5	100
	56-65	70	2.6	21	1.3	10	2.0	0	0.2	100
Sweden	16-25	4	1.0	17	1.5	40	2.0	40	1.4	100
	26-35	5	0.8	14	1.5	39	2.5	42	2.9	100
	36-45	7	0.7	20	1.4	42	1.3	32	1.1	100
	46-55	8	1.4	22	1.7	42	1.9	28	2.5	100
	56-65	16	2.5	33	1.8	35	2.1	16	2.4	100
Switzerland (French)	16-25	10	2.5	31	3.6	43	4.6	15	2.2	100
	26-35	11	1.7	29	2.1	46	2.8	13	2.1	100
	36-45	22	2.4	33	3.1	36	2.9	9	1.4	100
	46-55	21	3.7	35	2.6	36	3.5	8	1.6	100
	56-65	28	3.6	43	5.2	27	3.7	2	1.0	100
Switzerland (German)	16-25	7	1.7	35	3.4	43	5.1	14	3.3	100
	26-35	17	2.7	27	2.6	45	2.8	12	1.5	100
	36-45	24	2.1	34	3.0	32	3.2	9	1.4	100
	46-55	19	3.2	42	3.6	35	2.5	4	1.3	100
	56-65	30	5.1	46	4.4	20	3.1	4	1.6	100
United States	16-25*	24	2.3	31	2.6	33	2.7	13	2.0	100
	26-35	20	1.9	23	1.6	36	1.8	22	1.6	100
	36-45	19	1.9	21	2.1	30	2.9	29	2.0	100
	46-55	18	1.6	26	2.1	32	1.9	24	2.9	100
	56-65	24	1.9	31	2.7	31	2.6	15	2.2	100

* The proficiency of United States' post-secondary students has been underestimated due to a sampling anomaly.

s.e. Standard error of the estimate. The reported sample estimate can be said to be within 2 standard errors of the true population value with 95% confidence.

Table A6.1 (continued) **Literacy level by age group and country**

Country		Level 1		Level 2		Level 3		Level 4/5		Total
		%	se	%	se	%	se	%	se	%
Document literacy										
Canada	16-25	10	1.3	22	3.6	36	2.1	31	4.7	100
	26-35	13	2.9	25	3.3	34	5	27	3.5	100
	36-45	14	3.4	22	2.1	37	3.4	27	3.8	100
	46-55	23	4.7	31	3.4	24	6.8	22	10.7	100
	56-65	44	6.9	24	4.9	24	9.2	9	6.6	100
Germany	16-25	5	1.4	29	3.5	43	4.9	23	3.7	100
	26-35	6	1.2	29	2.9	40	2.2	25	1.9	100
	36-45	10	1.5	31	1.9	39	2.3	21	1.5	100
	46-55	7	1.3	35	4.3	43	3.4	15	2.5	100
	56-65	18	1.9	41	3.3	33	2.1	9	2	100
Great Britain	16-25	18	1.8	27	1.9	34	2.4	22	2	100
	26-35	19	1.6	25	2.2	32	1.8	23	1.6	100
	36-45	19	1.7	24	2.3	32	2	24	1.8	100
	46-55	24	2.3	28	1.7	31	2.9	16	1.6	100
	56-65	40	2.4	33	2.3	21	1.3	6	1	100
Netherlands	16-25	6	1.8	17	1.9	51	3	26	2.5	100
	26-35	6	1.3	19	1.5	46	2	29	1.5	100
	36-45	9	1.3	24	1.6	50	2.1	17	1.6	100
	46-55	13	1.7	36	2	38	2.4	14	1.8	100
	56-65	23	2.9	41	3.3	30	2.6	7	1.1	100
Poland	16-25	32	2.1	33	1.8	26	1.8	8	0.9	100
	26-35	39	2.2	34	1.6	20	1.4	7	0.9	100
	36-45	43	2	34	1.2	18	1.8	6	1	100
	46-55	56	2.4	27	2.5	13	2	4	0.8	100
	56-65	70	2.3	21	1.6	8	1.1	1	0.8	100
Sweden	16-25	3	0.8	17	1.9	40	1.5	41	1.6	100
	26-35	4	0.9	10	1.2	38	2.7	48	3.7	100
	36-45	7	0.5	18	1.6	40	1.7	35	1.6	100
	46-55	7	1	20	1.8	43	2.5	30	2.1	100
	56-65	12	1.8	33	1.8	36	2.3	19	2.4	100
Switzerland (French)	16-25	9	2	25	2.4	40	3.9	26	3.8	100
	26-35	11	1.7	22	2	45	3.2	22	2.5	100
	36-45	19	2.6	33	3	34	3.7	14	2	100
	46-55	18	3.3	30	3.8	42	3.9	10	2	100
	56-65	27	3.6	38	5	30	4.3	5	1.4	100
Switzerland (German)	16-25	7	1.9	26	4.2	41	3.7	26	3.2	100
	26-35	17	2.6	21	2.3	39	3.2	23	2.5	100
	36-45	21	1.9	30	2.3	36	2.4	12	1.6	100
	46-55	21	3	34	3.3	35	2.4	10	1.6	100
	56-65	23	4.1	40	3.1	31	4.7	7	1.9	100
United States	16-25*	25	2.2	31	2.8	28	3	16	2.4	100
	26-35	22	1.5	23	2.2	35	2.5	21	1.6	100
	36-45	24	1.9	20	1.5	31	2.3	25	1.8	100
	46-55	21	2.1	28	2.8	33	2.1	17	2.3	100
	56-65	29	1.9	33	2.4	26	1.8	12	1.7	100

*The proficiency of United States' post-secondary students has been underestimated due to a sampling anomoly

s.e. Standard error of the estimate. The reported sample estimate can be said to be within 2 standard errors of the true population value with 95% confidence.

Table A6.1 (continued) **Literacy level by age group and country**

Country		Level 1		Level 2		Level 3		Level 4/5		Total
		%	se	%	se	%	se	%	se	%
Quantitative literacy										
Canada	16-25	10	1	29	4.5	45	2.3	17	3.2	100
	26-35	12	3.4	26	5	35	9	27	3.9	100
	36-45	12	3	22	2.5	36	4.5	30	2.9	100
	46-55	24	3.5	32	6.8	25	4.1	19	8.2	100
	56-65	40	7.9	22	3.7	31	10.7	7	2.6	100
Germany	16-25	4	0.9	26	3.7	47	2.9	22	3.1	100
	26-35	5	0.9	23	1.8	43	2	29	2	100
	36-45	7	1.3	23	2.6	44	2.5	26	2	100
	46-55	7	1.5	27	3.9	41	3.2	25	3.2	100
	56-65	11	1.2	35	3.2	41	3.7	13	2.2	100
Great Britain	16-25	22	1.7	29	2.5	33	2.4	16	2.1	100
	26-35	20	1.7	28	2	30	2	23	1.6	100
	36-45	19	1.2	24	2	32	1.9	25	1.5	100
	46-55	24	2.3	26	2	33	2.8	17	1.7	100
	56-65	35	2.3	34	2.5	23	1.8	8	0.9	100
Netherlands	16-25	8	2	21	2.4	50	3.1	21	2.4	100
	26-35	7	1.1	20	1.2	45	2	28	1.6	100
	36-45	10	1.4	25	1.8	46	1.8	19	1.4	100
	46-55	13	1.8	31	1.9	40	2	16	1.5	100
	56-65	18	2.7	36	3.3	37	2.5	9	1.4	100
Poland	16-25	30	2.6	33	2.4	31	1.7	7	0.9	100
	26-35	33	2.4	33	2.3	26	0.7	9	1	100
	36-45	36	1.8	32	1.4	23	2.2	8	1.1	100
	46-55	48	2	27	2	20	1.3	6	0.9	100
	56-65	61	2.6	21	2.5	16	1.4	2	1.1	100
Sweden	16-25	5	0.7	18	1.8	39	2.2	38	1.9	100
	26-35	4	1	14	1.9	36	1.4	45	2.7	100
	36-45	7	0.6	16	1.4	41	1.5	35	1.5	100
	46-55	6	1.1	20	1.5	40	1.9	34	1.9	100
	56-65	13	2.1	27	2.7	38	2.5	23	2.3	100
Switzerland (French)	16-25	6	2	21	2.8	47	4.5	25	3.8	100
	26-35	9	1.6	21	2.5	48	3.6	23	2.3	100
	36-45	17	2.5	25	3.1	36	3	22	1.7	100
	46-55	16	3	23	2.4	43	3.8	18	2.4	100
	56-65	19	4.1	36	5.2	34	4	11	2	100
Switzerland (German)	16-25	7	2.3	22	4.2	48	3.2	23	2.9	100
	26-35	13	2.5	21	2.1	41	3	25	2.4	100
	36-45	19	2.1	26	3.1	38	3.3	17	2.3	100
	46-55	15	2.2	29	3.1	41	2.7	16	1.9	100
	56-65	16	3.5	38	4.7	36	5.5	11	2.8	100
United States	16-25*	27	1.9	31	2.5	29	2.8	13	2.3	100
	26-35	20	1.5	21	1.5	36	2.2	23	1.7	100
	36-45	18	2	23	2.4	27	2.5	32	1.5	100
	46-55	19	2	25	2.3	32	2	24	1.9	100
	56-65	22	1.4	30	2.6	32	2.1	16	2.2	100

* The proficiency of United States' post-secondary students has been underestimated due to a sampling anomaly.

s.e. Standard error of the estimate. The reported sample estimate can be said to be within 2 standard errors of the true population value with 95% confidence.

Table A6.2 **Literacy level by highest level of educational attainment (ISCED) and country**

Country		Level 1		Level 2		Level 3		Level 4/ 5		Total
		%	s.e.	%	s.e.	%	s.e.	%	s.e.	%
Prose literacy										
Canada	Primary education or lower	68	6.6	22	3.5	10	6.8	1	0.4	100
	Second level, 1st stage	22	2.2	37	3.2	33	4	8	2.5	100
	Second level, 2nd stage	10	3.2	29	4.5	41	5.6	19	0.9	100
	Third level, non-university	4	1.8	21	6.1	47	7.7	28	5	100
	Third level, university	0	0.3	11	6	30	5.7	59	9.4	100
Germany	Primary education or lower	68	12.7	15	8.1	18	8.9	0	0	100
	Second level, 1st stage	18	1.1	39	1.4	36	1.8	8	1.5	100
	Second level, 2nd stage	8	1.7	34	3.8	44	4.4	14	1.6	100
	Third level, non-university	4	1.8	14	3.1	49	5.5	33	3.7	100
	Third level, university	4	1	17	3.7	39	4.2	40	2.8	100
Great Britain	Primary education or lower	54	4.2	25	4	16	3.1	5	2.3	100
	Second level, 1st stage	28	1.5	36	1.6	27	1.6	9	0.7	100
	Second level, 2nd stage	12	1.6	30	2.1	37	2	21	2	100
	Third level, non-university	4	1.3	22	2.7	49	2.8	26	2.6	100
	Third level, university	3	0.9	13	1.7	38	2.4	46	2.3	100
Netherlands	Primary education or lower	38	3	42	3.1	17	2.4	3	0.8	100
	Second level, 1st stage	12	0.9	45	1.7	38	1.4	5	0.7	100
	Second level, 2nd stage	3	0.6	23	1.4	55	1.6	19	1.3	100
	Third level, non-university	..	..	..	..	..	..	..	..	..
	Third level, university	1	0.4	12	1.4	52	2.3	35	2.1	100
Poland	Primary education or lower	75	1.3	19	1	6	0.7	0	0.2	100
	Second level, 1st stage	42	1.6	40	1.8	16	1.2	2	0.4	100
	Second level, 2nd stage	25	1.9	44	2.5	28	1.8	2	0.7	100
	Third level, non-university	12	2.4	39	3.7	41	2.6	9	1.9	100
	Third level, university	11	3.2	30	2.8	42	5.1	16	3.3	100
Sweden	Primary education or lower	25	2.6	43	2.4	25	2.7	8	1.1	100
	Second level, 1st stage	7	1	21	1.4	47	3	25	2.7	100
	Second level, 2nd stage	6	0.7	20	1.1	43	1.6	31	1.2	100
	Third level, non-university	1	0.8	9	1.4	43	2.4	46	2.8	100
	Third level, university	1	0.5	6	1.7	32	3.9	61	4.6	100
Switzerland (French)	Primary education or lower	49	8.3	35	7.2	15	5.6	2	1.6	100
	Second level, 1st stage	29	4.6	52	5	20	5	0	0	100
	Second level, 2nd stage	11	1.8	36	2.3	44	2.9	9	0.8	100
	Third level, non-university	7	1.9	26	3.4	57	4.6	11	2.8	100
	Third level, university	5	2.1	13	2.7	49	4.2	32	3.5	100
Switzerland (German)	Primary education or lower	66	9.3	29	9.7	5	3.6	0	0	100
	Second level, 1st stage	34	4.9	43	5.4	19	4.6	4	2.1	100
	Second level, 2nd stage	11	1.1	40	1.9	39	1.9	10	1.1	100
	Third level, non-university	7	2.9	30	4.6	54	4.9	9	1.7	100
	Third level, university	7	2.4	21	5.6	47	4.9	26	5.7	100
United States	Primary education or lower	69	4.3	20	3.5	9	2	2	0.9	100
	Second level, 1st stage	45	4	30	3.5	22	2.4	3	0.7	100
	Second level, 2nd stage	17	1.6	34	2	35	2.1	14	1.8	100
	Third level, non-university	10	1.8	25	2.6	40	3.5	26	3.2	100
	Third level, university	5	1.2	12	0.9	36	2.4	48	2.7	100

s.e. Standard error of the estimate. The reported sample estimate can be said to be within 2 standard errors of the true population value with 95% confidence.

Table A6.2 (continued) Literacy level by highest level of educational attainment (ISCED) and country

Country		Level 1		Level 2		Level 3		Level 4/ 5		Total
		%	s.e.	%	s.e.	%	s.e.	%	s.e.	%
Document literacy										
Canada	Primary education or lower	74	7.6	15	3.4	10	6.8	1	0.8	100
	Second level, 1st stage	23	1.7	40	3.4	26	4.2	10	3.2	100
	Second level, 2nd stage	10	3.2	28	2.7	37	4.8	24	5.5	100
	Third level, non-university	4	2.7	18	2.8	39	5.9	39	6.9	100
	Third level, university	3	3.5	10	2.3	39	5.2	48	4.9	100
Germany	Primary education or lower	56	11.1	30	16.3	14	8.1	0	0	100
	Second level, 1st stage	10	0.9	38	1.5	39	2	12	1.8	100
	Second level, 2nd stage	5	1.1	27	3.3	43	2.7	25	1.6	100
	Third level, non-university	5	2.8	20	4.9	48	8.1	27	6	100
	Third level, university	1	0.7	18	3.3	35	3.4	46	3.5	100
Great Britain	Primary education or lower	51	4.6	25	3.9	17	3.4	7	2.3	100
	Second level, 1st stage	30	1.5	32	1.1	27	1.4	11	1	100
	Second level, 2nd stage	13	1.5	26	1.9	35	2.5	26	2.7	100
	Third level, non-university	7	1.6	21	2.3	45	3.2	27	2.7	100
	Third level, university	3	1	12	1.4	35	2.7	50	2.7	100
Netherlands	Primary education or lower	36	3.2	39	3.3	19	2.3	6	1.2	100
	Second level, 1st stage	11	1	37	1.7	43	1.6	9	0.8	100
	Second level, 2nd stage	3	0.7	18	1.1	52	1.6	26	1.6	100
	Third level, non-university	..	..	..	..	..	..	..	..	..
	Third level, university	1	0.4	14	1.3	50	2.2	35	1.8	100
Poland	Primary education or lower	75	1.5	19	1.2	5	0.6	1	0.4	100
	Second level, 1st stage	47	1	34	1.4	15	1.2	4	0.5	100
	Second level, 2nd stage	28	1.8	38	1.9	27	2.3	7	1	100
	Third level, non-university	16	3.5	35	3.6	36	2.7	12	2.4	100
	Third level, university	16	3.3	30	4.3	33	5.4	22	3	100
Sweden	Primary education or lower	23	2.3	38	2.1	33	2.9	6	1.1	100
	Second level, 1st stage	7	1.3	17	2.2	46	2.1	31	2.1	100
	Second level, 2nd stage	4	0.5	19	0.7	42	1	35	0.8	100
	Third level, non-university	1	0.5	11	2	38	3	50	1.9	100
	Third level, university	1	0.5	8	2.8	30	2.4	61	3.7	100
Switzerland (French)	Primary education or lower	42	7.4	40	6.8	16	4.2	2	2.1	100
	Second level, 1st stage	31	5.3	47	4.8	20	4.3	2	1.5	100
	Second level, 2nd stage	9	1.5	31	1.9	45	1.8	15	1.7	100
	Third level, non-university	2	1	20	3.8	48	4.9	31	4.7	100
	Third level, university	5	2.1	7	1.7	48	4.1	40	3.8	100
Switzerland (German)	Primary education or lower	73	10.8	17	8.2	11	5.7	0	0	100
	Second level, 1st stage	32	4.8	40	5.1	18	3.6	10	2.6	100
	Second level, 2nd stage	10	1.1	31	1.5	43	1.8	17	1.6	100
	Third level, non-university	5	1.9	25	4	49	5.2	21	3.7	100
	Third level, university	7	1.8	16	5	39	5.1	38	4.3	100
United States	Primary education or lower	74	3.8	19	3.6	6	1.6	1	0.6	100
	Second level, 1st stage	45	4.5	28	3	21	3.5	6	1.7	100
	Second level, 2nd stage	21	1.2	34	1.6	32	1.8	13	1.3	100
	Third level, non-university	12	1.8	25	2.7	39	2.5	24	1.9	100
	Third level, university	7	1.8	13	1.3	39	1.8	41	2.6	100

s.e. Standard error of the estimate. The reported sample estimate can be said to be within 2 standard errors of the true population value with 95% confidence.

Table A6.2 (continued) Literacy level by highest level of educational attainment (ISCED) and country

Country		Level 1		Level 2		Level 3		Level 4/ 5		Total
		%	s.e.	%	s.e.	%	s.e.	%	s.e.	%
Quantitative literacy										
Canada	Primary education or lower	69	7.3	19	4.1	11	6.5	1	0.3	100
	Second level, 1st stage	23	2.3	42	5.0	28	3.8	8	1.9	100
	Second level, 2nd stage	9	4.7	32	5.0	43	6.7	17	2.6	100
	Third level, non-university	4	3.2	21	4.4	49	4.2	26	4.3	100
	Third level, university	2	3.5	4	1.2	29	9.4	64	11.3	100
Germany	Primary education or lower	43	16.3	21	11.4	29	11.2	8	8.3	100
	Second level, 1st stage	8	0.5	31	2.2	44	1.4	17	1.5	100
	Second level, 2nd stage	4	1.2	21	2.7	49	1.5	26	2.7	100
	Third level, non-university	3	1.4	11	4.1	59	6.6	27	4.6	100
	Third level, university	2	0.8	13	3.5	29	2.3	56	2.5	100
Great Britain	Primary education or lower	50	3.8	33	3.9	14	2.8	3	2.0	100
	Second level, 1st stage	30	1.2	32	1.5	29	1.4	9	0.8	100
	Second level, 2nd stage	13	1.6	27	2.3	36	2.0	24	2.4	100
	Third level, non-university	7	1.6	22	2.4	41	2.7	29	2.9	100
	Third level, university	3	0.8	12	1.4	29	2.6	56	2.7	100
Netherlands	Primary education or lower	35	3.4	36	3.2	24	3.0	5	1.1	100
	Second level, 1st stage	12	1.2	36	1.8	42	1.6	11	0.9	100
	Second level, 2nd stage	3	0.6	22	1.5	52	1.5	23	1.6	100
	Third level, non-university	..	..	..	..	..	..	..	..	..
	Third level, university	2	0.6	10	1.2	49	2.3	39	2.6	100
Poland	Primary education or lower	69	1.3	21	1.1	8	0.6	1	0.4	100
	Second level, 1st stage	39	1.2	34	1.8	22	1.3	4	0.6	100
	Second level, 2nd stage	21	1.6	36	1.3	33	1.2	10	1.3	100
	Third level, non-university	16	2.6	26	2.2	48	3.5	11	2.0	100
	Third level, university	9	2.1	26	3.9	39	4.1	26	3.3	100
Sweden	Primary education or lower	22	2.1	32	1.6	35	2.2	11	1.5	100
	Second level, 1st stage	7	1.2	21	2.7	41	2.4	31	2.2	100
	Second level, 2nd stage	5	0.6	18	0.6	42	1.1	35	1.2	100
	Third level, non-university	1	0.4	15	2.2	39	2.7	46	2.8	100
	Third level, university	1	0.7	6	2.2	29	2.2	64	3.4	100
Switzerland (French)	Primary education or lower	40	7.8	37	6.0	22	4.2	0	0.3	100
	Second level, 1st stage	23	3.2	44	3.2	29	4.4	4	2.3	100
	Second level, 2nd stage	6	0.9	24	2.1	48	1.9	22	1.7	100
	Third level, non-university	3	1.3	14	3.2	52	5.0	31	3.8	100
	Third level, university	4	1.7	9	1.9	45	4.6	41	5.0	100
Switzerland (German)	Primary education or lower	51	9.6	26	12.2	20	7.5	3	3.2	100
	Second level, 1st stage	22	4.8	44	5.3	21	3.8	13	4.1	100
	Second level, 2nd stage	7	1.0	27	1.7	47	2.1	19	1.8	100
	Third level, non-university	4	1.3	14	3.8	54	5.5	28	5.3	100
	Third level, university	7	1.9	18	4.3	36	4.0	39	5.0	100
United States	Primary education or lower	67	4.1	23	3.7	9	2.1	1	0.5	100
	Second level, 1st stage	45	4.3	23	2.8	28	4.1	4	1.7	100
	Second level, 2nd stage	18	1.1	34	1.7	33	1.5	14	1.4	100
	Third level, non-university	9	1.9	23	3.6	41	3.7	27	3.4	100
	Third level, university	5	1.1	11	1.1	32	2.0	52	2.2	100

s.e. Standard error of the estimate. The reported sample estimate can be said to be within 2 standard errors of the true population value with 95% confidence.

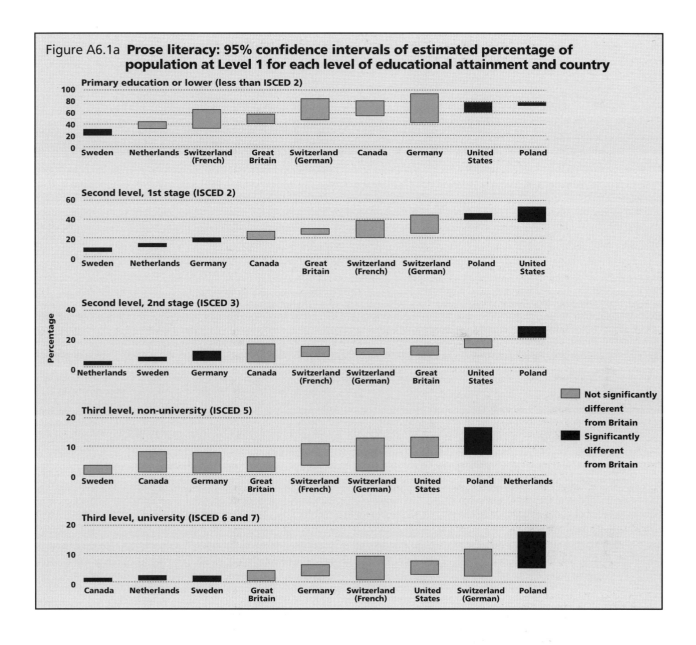

Figure A6.1a **Prose literacy: 95% confidence intervals of estimated percentage of population at Level 1 for each level of educational attainment and country**

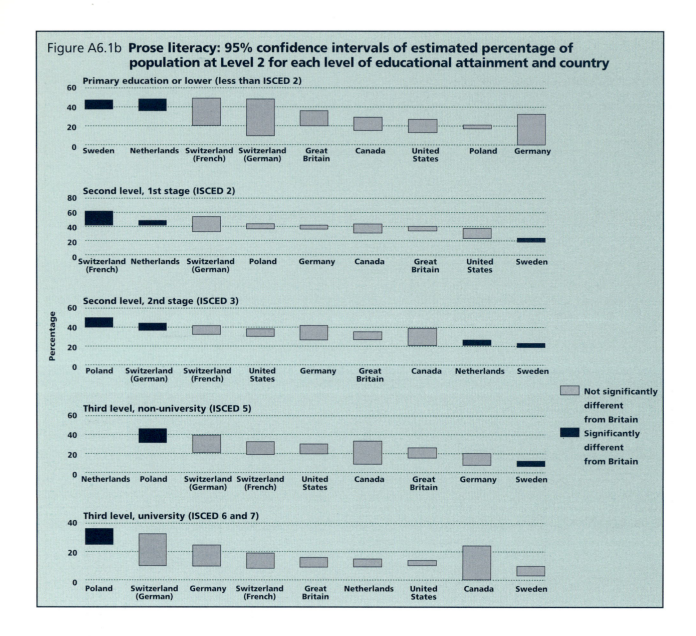

Figure A6.1b **Prose literacy: 95% confidence intervals of estimated percentage of population at Level 2 for each level of educational attainment and country**

Figure A6.1c **Prose literacy: 95% confidence intervals of estimated percentage of population at Level 3 for each level of educational attainment and country**

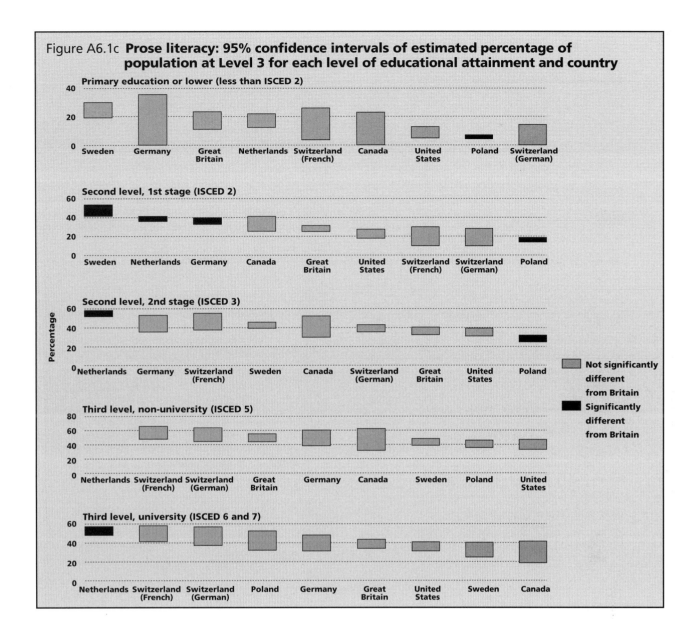

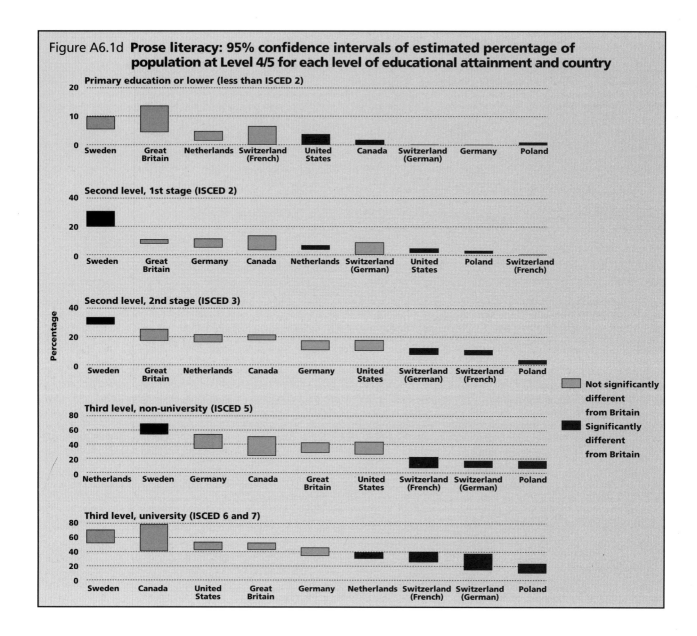

Figure A6.1d **Prose literacy: 95% confidence intervals of estimated percentage of population at Level 4/5 for each level of educational attainment and country**

Table A6.3 **Literacy level by economic activity status and country**

Country		Level 1	Level 2	Level 3	Level 4/5	Total
		%	%	%	%	%
Prose literacy						
Canada	Employed	12	25	38	26	100
	Unemployed	33	23	36	9	100
	Student	12	23	40	26	100
	Inactive	29	31	25	15	100
Germany	Employed	11	33	41	15	100
	Unemployed	26	34	28	12	100
	Student	5	22	44	29	100
	Inactive	20	39	34	7	100
Great Britain	Employed	15	30	35	20	100
	Unemployed	32	28	28	12	100
	Student	7	28	37	28	100
	Inactive	40	33	20	7	100
Netherlands	Employed	7	27	49	18	100
	Unemployed	20	29	43	8	100
	Student	7	19	50	24	100
	Inactive	19	42	32	8	100
Poland	Employed	39	36	21	4	100
	Unemployed	45	39	16	1	100
	Student	16	37	38	8	100
	Inactive	60	28	11	1	100
Sweden	Employed	6	19	41	34	100
	Unemployed	13	22	41	25	100
	Student	3	15	40	42	100
	Inactive	21	30	32	18	100
Switzerland (French)	Employed	14	34	42	11	100
	Unemployed	25	27	35	12	100
	Student	5	30	44	20	100
	Inactive	24	43	29	3	100
Switzerland (German)	Employed	15	38	38	9	100
	Unemployed	20	41	30	9	100
	Student	3	18	51	28	100
	Inactive	25	42	29	4	100
United States	Employed	15	26	34	25	100
	Unemployed	31	27	32	10	100
	Student	27	25	34	14	100
	Inactive	32	26	28	14	100

Table A6.3 (continued) **Literacy level by economic activity status and country**

Country		Level 1	Level 2	Level 3	Level 4/5	Total
		%	%	%	%	%
Document literacy						
Canada	Employed	12	24	35	30	100
	Unemployed	30	29	23	17	100
	Student	8	26	32	34	100
	Inactive	38	25	28	10	100
Germany	Employed	5	31	42	22	100
	Unemployed	18	41	26	15	100
	Student	6	18	37	39	100
	Inactive	15	38	39	8	100
Great Britain	Employed	16	26	34	23	100
	Unemployed	32	33	22	13	100
	Student	7	24	41	28	100
	Inactive	21	29	33	8	100
Netherlands	Employed	6	22	48	24	100
	Unemployed	17	26	47	11	100
	Student	5	14	47	34	100
	Inactive	21	38	33	8	100
Poland	Employed	41	31	21	7	100
	Unemployed	47	33	16	3	100
	Student	23	35	32	11	100
	Inactive	63	27	8	2	100
Sweden	Employed	5	17	41	38	100
	Unemployed	12	23	35	30	100
	Student	3	17	40	41	100
	Inactive	15	32	33	20	100
Switzerland (French)	Employed	12	30	41	17	100
	Unemployed	15	28	36	21	100
	Student	2	20	50	28	100
	Inactive	28	32	31	9	100
Switzerland (German)	Employed	14	31	38	17	100
	Unemployed	24	23	35	17	100
	Student	2	16	42	40	100
	Inactive	22	35	35	8	100
United States	Employed	18	26	34	23	100
	Unemployed	36	27	25	13	100
	Student	24	26	33	17	100
	Inactive	37	30	25	8	100

Table A6.3 (continued)   **Literacy level by economic activity status and country**

Country		Level 1	Level 2	Level 3	Level 4/5	Total
		%	%	%	%	%
Quantitative literacy						
Canada	Employed	11	25	36	27	100
	Unemployed	33	31	27	9	100
	Student	8	27	45	21	100
	Inactive	32	27	30	11	100
Germany	Employed	4	23	45	28	100
	Unemployed	10	33	40	16	100
	Student	4	20	38	38	100
	Inactive	11	34	41	14	100
Great Britain	Employed	16	27	33	23	100
	Unemployed	33	31	27	9	100
	Student	11	32	32	24	100
	Inactive	44	28	22	6	100
Netherlands	Employed	6	21	48	25	100
	Unemployed	18	26	44	11	100
	Student	6	19	50	25	100
	Inactive	20	37	34	9	100
Poland	Employed	35	31	26	9	100
	Unemployed	43	30	22	4	100
	Student	19	37	36	8	100
	Inactive	56	26	15	2	100
Sweden	Employed	5	17	40	38	100
	Unemployed	10	20	36	33	100
	Student	5	18	38	39	100
	Inactive	18	25	36	22	100
Switzerland (French)	Employed	9	25	46	21	100
	Unemployed	13	24	40	23	100
	Student	3	14	46	36	100
	Inactive	21	33	33	14	100
Switzerland (German)	Employed	10	26	42	21	100
	Unemployed	19	38	38	6	100
	Student	2	21	42	35	100
	Inactive	14	33	42	11	100
United States	Employed	16	24	32	27	100
	Unemployed	37	24	27	12	100
	Student	26	27	37	11	100
	Inactive	30	29	29	12	100

151

Table A6.4 (continued) Literacy level by occupation

Country		Level 1	Level 2	Level 3	Level 4/5	Total
		%	%	%	%	%
Prose literacy						
Canada	Manager/Professional	3	17	37	43	100
	Technician	4	26	26	43	100
	Clerks	6	28	51	15	100
	Sales/Service worker	11	29	35	25	100
	Skilled craft worker	30	23	33	14	100
	Machinery operator/Assembler	29	20	40	11	100
	Agriculture/Primary occupation	19	28	40	14	100
Germany	Manager/Professional	5	19	44	32	100
	Technician	4	23	49	24	100
	Clerks	10	39	39	13	100
	Sales/Service worker	10	37	36	16	100
	Skilled craft worker	14	36	43	7	100
	Machinery operator/Assembler	22	53	20	6	100
	Agriculture/Primary occupation	37	31	28	4	100
Great Britain	Manager/Professional	6	21	43	31	100
	Technician	10	20	40	30	100
	Clerks	12	29	37	23	100
	Sales/Service worker	14	35	33	19	100
	Skilled craft worker	25	38	31	6	100
	Machinery operator/Assembler	28	45	22	5	100
	Agriculture/Primary occupation	36	32	23	9	100
Netherlands	Manager/Professional	3	20	52	25	100
	Technician	3	20	54	23	100
	Clerks	6	24	53	17	100
	Sales/Service worker	9	30	44	18	100
	Skilled craft worker	10	45	38	7	100
	Machinery operator/Assembler	19	36	37	8	100
	Agriculture/Primary occupation	17	32	43	8	100
Poland	Manager/Professional	13	31	41	15	100
	Technician	23	45	28	4	100
	Clerks	25	43	28	3	100
	Sales/Service worker	30	43	22	4	100
	Skilled craft worker	47	39	14	0	100
	Machinery operator/Assembler	49	35	16	0	100
	Agriculture/Primary occupation	63	28	9	1	100
Sweden	Manager/Professional	2	12	38	47	100
	Technician	3	17	43	37	100
	Clerks	3	19	43	35	100
	Sales/Service worker	7	22	39	32	100
	Skilled craft worker	10	26	42	21	100
	Machinery operator/Assembler	8	28	41	23	100
	Agriculture/Primary occupation	12	30	39	19	100
Switzerland (French)	Manager/Professional	7	17	53	22	100
	Technician	8	30	49	14	100
	Clerks	3	39	46	12	100
	Sales/Service worker	27	46	25	3	100
	Skilled craft worker	25	36	38	1	100
	Machinery operator/Assembler	28	30	32	10	100
	Agriculture/Primary occupation	25	48	24	3	100
Switzerland (German)	Manager/Professional	5	31	50	13	100
	Technician	3	30	53	14	100
	Clerks	6	38	40	15	100
	Sales/Service worker	16	44	35	5	100
	Skilled craft worker	25	46	26	2	100
	Machinery operator/Assembler	40	36	24	0	100
	Agriculture/Primary occupation	33	44	20	2	100
United States	Manager/Professional	4	16	37	43	100
	Technician	2	16	47	34	100
	Clerks	7	30	42	21	100
	Sales/Service worker	24	26	32	17	100
	Skilled craft worker	29	38	25	7	100
	Machinery operator/Assembler	29	37	28	6	100
	Agriculture/Primary occupation	32	21	24	23	100

Table A6.4 (continued) **Literacy level by occupation**

Country		Level 1	Level 2	Level 3	Level 4/5	Total
		%	%	%	%	%
Document literacy						
Canada	Manager/Professional	3	15	32	50	100
	Technician	3	12	59	26	100
	Clerks	8	27	37	28	100
	Sales/Service worker	16	30	29	25	100
	Skilled craft worker	25	30	29	16	100
	Machinery operator/Assembler	28	31	26	15	100
	Agriculture/Primary occupation	18	31	33	18	100
Germany	Manager/Professional	1	20	36	42	100
	Technician	2	14	54	30	100
	Clerks	5	31	44	19	100
	Sales/Service worker	6	37	39	18	100
	Skilled craft worker	7	33	47	14	100
	Machinery operator/Assembler	12	48	32	8	100
	Agriculture/Primary occupation	19	39	29	13	100
Great Britain	Manager/Professional	6	19	38	36	100
	Technician	13	16	40	31	100
	Clerks	13	27	35	25	100
	Sales/Service worker	17	28	36	18	100
	Skilled craft worker	24	36	28	12	100
	Machinery operator/Assembler	22	42	28	8	100
	Agriculture/Primary occupation	37	27	27	9	100
Netherlands	Manager/Professional	2	17	53	28	100
	Technician	3	15	50	33	100
	Clerks	5	20	55	20	100
	Sales/Service worker	7	24	49	20	100
	Skilled craft worker	9	36	39	16	100
	Machinery operator/Assembler	13	33	36	18	100
	Agriculture/Primary occupation	16	24	44	16	100
Poland	Manager/Professional	19	28	34	18	100
	Technician	22	39	30	9	100
	Clerks	33	32	28	7	100
	Sales/Service worker	34	33	26	7	100
	Skilled craft worker	47	30	17	6	100
	Machinery operator/Assembler	58	27	13	2	100
	Agriculture/Primary occupation	61	29	9	1	100
Sweden	Manager/Professional	2	14	38	46	100
	Technician	3	15	42	41	100
	Clerks	2	16	41	41	100
	Sales/Service worker	6	21	41	31	100
	Skilled craft worker	8	17	45	30	100
	Machinery operator/Assembler	7	19	45	28	100
	Agriculture/Primary occupation	11	25	38	26	100
Switzerland (French)	Manager/Professional	5	16	49	30	100
	Technician	7	30	48	15	100
	Clerks	6	31	46	16	100
	Sales/Service worker	17	39	35	9	100
	Skilled craft worker	22	29	32	17	100
	Machinery operator/Assembler	28	35	23	14	100
	Agriculture/Primary occupation	20	45	28	7	100
Switzerland (German)	Manager/Professional	5	29	44	22	100
	Technician	4	22	48	25	100
	Clerks	7	32	42	19	100
	Sales/Service worker	20	38	36	6	100
	Skilled craft worker	22	37	33	8	100
	Machinery operator/Assembler	31	27	31	11	100
	Agriculture/Primary occupation	31	32	25	12	100
United States	Manager/Professional	5	15	41	39	100
	Technician	4	17	49	30	100
	Clerks	11	34	33	22	100
	Sales/Service worker	27	25	33	15	100
	Skilled craft worker	30	38	25	7	100
	Machinery operator/Assembler	35	32	26	7	100
	Agriculture/Primary occupation	36	12	27	24	100

Table A6.4 (continued) **Literacy level by occupation**

Country		Level 1	Level 2	Level 3	Level 4/5	Total
		%	%	%	%	%
Quantitative literacy						
Canada	Manager/Professional	2	15	36	46	100
	Technician	4	18	33	45	100
	Clerks	5	35	41	20	100
	Sales/Service worker	15	31	41	13	100
	Skilled craft worker	22	35	29	14	100
	Machinery operator/Assembler	29	29	34	9	100
	Agriculture/Primary occupation	21	25	36	18	100
Germany	Manager/Professional	2	14	37	47	100
	Technician	2	15	52	31	100
	Clerks	5	26	46	23	100
	Sales/Service worker	5	25	45	25	100
	Skilled craft worker	3	24	48	25	100
	Machinery operator/Assembler	11	41	36	12	100
	Agriculture/Primary occupation	18	27	39	17	100
Great Britain	Manager/Professional	6	17	38	39	100
	Technician	11	19	40	31	100
	Clerks	15	27	36	21	100
	Sales/Service worker	21	31	34	13	100
	Skilled craft worker	21	38	27	14	100
	Machinery operator/Assembler	20	42	26	12	100
	Agriculture/Primary occupation	37	32	23	8	100
Netherlands	Manager/Professional	2	15	49	34	100
	Technician	3	17	51	29	100
	Clerks	5	27	52	17	100
	Sales/Service worker	8	24	47	21	100
	Skilled craft worker	10	32	44	14	100
	Machinery operator/Assembler	13	25	42	20	100
	Agriculture/Primary occupation	18	27	44	10	100
Poland	Manager/Professional	12	26	38	25	100
	Technician	19	33	36	13	100
	Clerks	28	32	30	11	100
	Sales/Service worker	28	37	28	7	100
	Skilled craft worker	42	29	24	5	100
	Machinery operator/Assembler	43	31	20	7	100
	Agriculture/Primary occupation	54	28	15	2	100
Sweden	Manager/Professional	2	15	37	46	100
	Technician	4	15	42	40	100
	Clerks	4	15	42	39	100
	Sales/Service worker	7	21	40	31	100
	Skilled craft worker	6	20	44	30	100
	Machinery operator/Assembler	8	16	42	34	100
	Agriculture/Primary occupation	8	26	39	26	100
Switzerland (French)	Manager/Professional	4	11	45	41	100
	Technician	4	19	58	20	100
	Clerks	3	25	52	20	100
	Sales/Service worker	20	36	34	10	100
	Skilled craft worker	12	28	40	19	100
	Machinery operator/Assembler	27	31	33	8	100
	Agriculture/Primary occupation	19	39	37	6	100
Switzerland (German)	Manager/Professional	4	16	50	30	100
	Technician	3	20	49	28	100
	Clerks	8	26	45	20	100
	Sales/Service worker	12	39	38	11	100
	Skilled craft worker	12	37	40	13	100
	Machinery operator/Assembler	28	25	40	8	100
	Agriculture/Primary occupation	26	33	27	14	100
United States	Manager/Professional	4	14	37	46	100
	Technician	2	11	44	42	100
	Clerks	11	32	35	22	100
	Sales/Service worker	25	28	29	17	100
	Skilled craft worker	29	32	29	11	100
	Machinery operator/Assembler	30	31	28	11	100
	Agriculture/Primary occupation	34	10	42	14	100

Table A6.5 **Percentage who reported engaging in each of several workplace reading activities at least once a week by occupation**

People who had worked in the 12 months prior to interview

Country		Letters or memos	Reports, articles, magazines or journals	Manuals or reference books, including catalogues	Diagrams or schematics	Bills, invoices, spreadsheets or budget tables	Directions or instructions for medicines, recipes or other products
		Percentage engaging in each activity					
Canada	Manager/ Professional	93	84	82	52	65	45
	Technician	77	70	48	28	45	28
	Clerks	84	53	46	15	52	17
	Sales/ Service worker	58	41	32	15	44	35
	Skilled craft worker	42	30	37	48	33	23
	Machinery operator/ Assembler	56	41	32	33	33	21
	Agriculture/ Primary occupation	42	32	30	23	38	25
Germany	Manager/ Professional	92	86	68	62	58	26
	Technician	93	89	67	54	69	39
	Clerks	96	79	64	39	76	26
	Sales/ Service worker	74	63	59	33	64	38
	Skilled craft worker	69	43	59	73	54	32
	Machinery operator/ Assembler	68	44	51	50	51	40
	Agriculture/ Primary occupation	45	31	35	22	40	30
Great Britain	Manager/ Professional	94	88	78	53	66	40
	Technician	91	83	72	44	50	45
	Clerks	87	62	57	22	62	18
	Sales/ Service worker	55	42	38	20	32	37
	Skilled craft worker	62	40	50	58	35	20
	Machinery operator/ Assembler	58	40	39	41	28	26
	Agriculture/ Primary occupation	40	24	17	16	21	17
Netherlands	Manager/ Professional	87	87	71	55	60	26
	Technician	82	74	62	44	51	34
	Clerks	80	59	51	37	50	18
	Sales/ Service worker	44	43	32	19	37	29
	Skilled craft worker	38	37	51	40	23	16
	Machinery operator/ Assembler	54	43	35	38	18	16
	Agriculture/ Primary occupation	27	27	15	18	17	11
Poland	Manager/ Professional	72	79	71	42	48	51
	Technician	65	61	52	34	42	39
	Clerks	61	44	35	13	64	27
	Sales/ Service worker	28	20	24	4	53	32
	Skilled craft worker	17	11	13	29	8	15
	Machinery operator/ Assembler	24	12	10	20	23	11
	Agriculture/ Primary occupation	8	11	8	4	9	10
Sweden	Manager/ Professional	92	92	82	77	62	..
	Technician	88	89	77	70	64	..
	Clerks	90	84	79	53	66	..
	Sales/ Service worker	75	77	69	54	64	..
	Skilled craft worker	57	62	62	50	42	..
	Machinery operator/ Assembler	55	57	45	47	43	..
	Agriculture/ Primary occupation	50	53	47	38	39	..
Switzerland (French)	Manager/ Professional	86	88	55	50	78	14
	Technician	80	78	55	42	55	29
	Clerks	83	75	39	22	68	9
	Sales/ Service worker	57	52	42	20	47	22
	Skilled craft worker	54	62	55	56	47	21
	Machinery operator/ Assembler	62	58	38	31	35	18
	Agriculture/ Primary occupation	64	64	52	25	45	25
Switzerland (German)	Manager/ Professional	95	90	66	47	70	18
	Technician	94	82	65	41	62	28
	Clerks	91	74	57	25	73	8
	Sales/ Service worker	69	59	41	18	62	16
	Skilled craft worker	69	56	63	32	55	15
	Machinery operator/ Assembler	54	39	27	12	47	17
	Agriculture/ Primary occupation	56	49	36	15	42	7
United States	Manager/ Professional	92	85	83	52	63	42
	Technician	83	75	81	64	33	48
	Clerks	88	68	60	25	59	28
	Sales/ Service worker	56	44	48	22	41	35
	Skilled craft worker	53	39	53	58	35	26
	Machinery operator/ Assembler	44	29	42	28	21	21
	Agriculture/ Primary occupation	26	35	13	9	25	20

155

Table A6.6 **Percentage who reported engaging in each of several workplace writing and mathematical activities at least once a week by occupation**
People who had worked in the 12 months prior to interview

Country		Writing tasks				Mathematics tasks	
		Letters or memos	Forms, bills, invoices or budgets	Reports or articles	Estimates or technical specifications	Meaure or estimate the size or weight of objects	Calculate prices, costs or budgets
		Percentage engaging in each activity					
Canada	Manager/ Professional	84	58	65	43	48	30
	Technician	49	37	37	24	31	40
	Clerks	64	62	34	14	41	52
	Sales/ Service worker	41	46	27	15	42	56
	Skilled craft worker	33	29	28	36	66	41
	Machinery operator/ Assembler	36	41	37	28	55	30
	Agriculture/ Primary occupation	28	34	15	12	48	38
Germany	Manager/ Professional	91	60	64	27	34	48
	Technician	87	70	70	34	29	49
	Clerks	84	73	60	34	18	63
	Sales/ Service worker	68	60	36	14	11	49
	Skilled craft worker	58	43	34	35	49	30
	Machinery operator/ Assembler	58	56	27	17	21	14
	Agriculture/ Primary occupation	29	24	10	8	15	22
Great Britain	Manager/ Professional	84	66	61	32	45	61
	Technician	77	55	48	32	50	52
	Clerks	68	61	30	16	32	51
	Sales/ Service worker	30	36	24	8	37	43
	Skilled craft worker	38	49	28	29	68	42
	Machinery operator/ Assembler	30	35	24	10	50	17
	Agriculture/ Primary occupation	22	28	11	6	29	18
Netherlands	Manager/ Professional	70	32	52	44	50	58
	Technician	72	32	48	30	41	50
	Clerks	65	30	28	22	35	49
	Sales/ Service worker	35	30	25	18	39	52
	Skilled craft worker	27	16	24	40	61	41
	Machinery operator/ Assembler	30	11	19	32	53	31
	Agriculture/ Primary occupation	20	9	11	12	33	18
Poland	Manager/ Professional	66	49	33	20	51	43
	Technician	59	48	33	15	51	35
	Clerks	51	69	36	11	41	46
	Sales/ Service worker	21	47	21	8	66	53
	Skilled craft worker	15	10	8	6	58	10
	Machinery operator/ Assembler	21	24	14	4	42	16
	Agriculture/ Primary occupation	7	7	4	2	31	16
Sweden	Manager/ Professional	88	53	65	39	57	52
	Technician	86	49	59	23	36	53
	Clerks	83	61	56	23	30	47
	Sales/ Service worker	66	52	50	21	54	66
	Skilled craft worker	48	37	43	43	78	31
	Machinery operator/ Assembler	49	32	35	22	57	34
	Agriculture/ Primary occupation	48	31	32	19	48	30
Switzerland (French)	Manager/ Professional	79	78	60	31	33	56
	Technician	65	46	52	29	40	49
	Clerks	80	65	41	18	20	45
	Sales/ Service worker	43	35	28	10	25	47
	Skilled craft worker	42	37	46	43	70	37
	Machinery operator/ Assembler	48	48	42	8	44	33
	Agriculture/ Primary occupation	38	30	36	12	34	40
Switzerland (German)	Manager/ Professional	92	67	65	38	44	53
	Technician	90	65	53	29	53	41
	Clerks	91	64	44	21	24	34
	Sales/ Service worker	68	62	27	19	25	33
	Skilled craft worker	70	38	27	28	75	29
	Machinery operator/ Assembler	50	49	22	1	56	16
	Agriculture/ Primary occupation	55	37	27	12	35	24
United States	Manager/ Professional	83	64	65	39	45	64
	Technician	74	44	49	47	61	35
	Clerks	73	58	49	19	33	54
	Sales/ Service worker	44	44	29	20	42	54
	Skilled craft worker	41	48	35	46	74	50
	Machinery operator/ Assembler	21	34	23	18	58	23
	Agriculture/ Primary occupation	20	24	6	9	50	32

Appendices

Appendix A
Survey design and response

This appendix describes the British Adult Literacy survey which was conducted as part of an international programme of surveys known as the International Adult Literacy Survey (IALS). A description of the sample design and response for other countries participating in the survey will be published in the IALS technical report.

1. Survey design
The survey aimed to interview a representative sample of adults aged 16-65 inclusive resident in private households in Britain. The survey was restricted to those in this age range because of the relevance of the skills being measured to the labour market as well as known difficulties associated with administering such assessments to older people. Because of the level of respondent burden expected for some respondents, the sensitive nature of the survey and in order to prevent contamination of the test between respondents in the same household or living at the same address the survey was designed so that only one person per household and only one household per address would be interviewed.

2. Sample design
A sample of addresses was selected from the small users version of the postcode address file (PAF). A multi-stage sample design was used with postcode sectors as the primary sampling units (PSUs). The design involves both stratification and clustering. Postcode sectors were stratified by region and then by selected economic indicators from the 1981 Census. Within each major stratum, sectors were reordered using the census variables (SEG and proportion of households with no car available) and split into equal sized minor strata depending on the number of selections required from the region. Each minor strata is then divided into as many sections as there are selections required. One section is then picked at random and a PSU selected at random from the selected section with probability proportionate to size.

3. Sampling in the field
The sample design required that only one household per address and only one person per household be interviewed. Where more than one household was found at an address, the interviewer listed all the households in a set order and selected one household to interview using a standard procedure. Where more than one member of the household was eligible for the survey, the interviewer selected the person to be interviewed according to set procedures. All persons eligible for the survey were listed in descending age order. The interviewer used a Kish reference grid which identified the sampled individual for each address and for each possible number of eligible household members.

As a large proportion of households do not contain anyone aged 16-65 the initial sample selection needed to accommodate the expected loss due to ineligibility on grounds of age in addition to ineligibilty because the address did not contain a private household. A total of 8,290 addresses were selected of which almost one third were not eligible for survey. One in five of all sampled addresses did not contain anyone in the required age range.

Table A.1 **Sample of addresses and response**

	N	%
Selected addresses	8290	
Ineligible addresses	2687	
No one aged 16-65	1655	
Demolished or derelict	80	
Used solely for business purposes	222	
Temporary accommodation only	90	
Vacant	406	
Address not traced	179	
Institution	18	
Under construction	36	
Other ineligible	1	
Total eligible sample of addresses	5603	
Co-operating respondents	3811	68.02
Refusal	891	15.90
Non-contact	380	6.78
Contact made - no interview	521	9.30
Total	5603	100.00

4. Response to the survey

From the sample of addresses 5,603 contained eligible households. Overall, 68% of eligible households co-operated with the survey. This response was disappointing compared with other surveys conducted in this country but compares very favourably with response rates for the same survey in other countries. It is difficult to make exact comparisons with some countries as they trailed the IALS onto an existing survey such as the Labour Force Survey. Response to the survey was lowest in the Netherlands at 44.8% however response to surveys in the Netherlands is generally low. Other countries which used a fresh sample as opposed to a follow on to an existing survey achieved better response rate but still low compared with other surveys, Sweden 60% , Switzerland 59.6% and 50.3% for the German and French speaking samples respectively and Germany 69%. Both Canada and the US used existing surveys to select the IALS sample. Thus the response rates include non-response to the original survey and non-response to the IALS. In Canada 73.2% of the IALS sample responded which gives an overall response rate of 67.4% when the Labour Force Survey response rate is taken into account. Similarly, in the US a composite response rate of 59.4% was achieved when response to the Current Population Survey (95%) and response to IALS (62.4%) are taken into account.[1]

(Table A.1)

5. Representativeness of the sample

The characteristics of the responding sample were compared with their distribution in the general population based on another survey. Males and young people were found to be under represented in the sample when compared to the Labour Force Survey (LFS) estimates for these characteristics. Because of the relatively small sample size of the survey and the need to weight and gross the data the age and education categories had to be collapsed to ensure a minimum number of cases in each weighting cell.

(Table A.2)

6. Data collection

Fieldwork was carried out between February and June 1996. The interview consisted of a background questionnaire which collected socio-demographic information as well as questions on literacy practices,

Table A.2 **Comparison of weighted and unweighted Adult Literacy Sample with Labour Force Survey by key characteristics**

Characteristic	Adult Literacy Survey - unweighted	Labour Force Survey	Adult Literacy Survey - weighted
Sex			
Male	45.4	50.3	50.3
Female	54.6	49.7	49.7
Age group			
16-24	12.3	17.6	17.6
25-34	24.9	24.9	22.5
35-44	23.2	21.0	23.3
45-54	19.6	19.7	19.3
55-65	20.0	16.9	17.3
Highest level of educational attainment			
Second level, 1st stage or lower	55.8	56.9	60.8
Second level, 2nd stage	19.1	19.3	19.9
Third level	25.2	23.3	19.2
Employment status			
Employed/self-employed	68.6	68.9	69.8
Unemployed	7.2	7.1	7.1
Full-time student	1.9	3.9	3.0
Inactive/Other	22.3	20.1	20.2

*LFS proportions are based on weighted Spring 1995 file which was the most recent file to contain ISCED level. Adult Literacy weighting took account of sample design and weighted to LFS cell proportion of education level by age group by sex.

a screening assessment to identify those with very limited literacy skills and the main assessment. The assessment design used a Balanced Incomplete Block system to spiral the assessment items into different combinations. The particular combination of tasks that a respondent was asked to do (represented by the Booklet number) was allocated based on the address number within the quota of 35 addresses.

7. Scoring of the booklets

The assessment booklets were scored by the interviewers using a computer assisted coding programme. To date Britain is one of only two countries to have the booklets scored by interviewers. The scoring guidelines for the assessment are very clear and straight forward for the vast majority of items. For most items the answer is either right or wrong and the scorer does not need to make a judgement about whether the answer provided by the respondent matched the answer in the guide. There are about 10 items where the scorer has to make a judgement as to whether the answer is sufficiently close to or equivalent to the model answer. Most of these items are where the respondent has to answer in their own words, make comparisons or describe a relationship. These items are the most difficult to score.

For all items, interviewers could telephone the office based team for assistance in scoring or for an adjudication when they were unsure about how to score a particular item. Alternatively, they were asked to mark any items they were unsure how to score with page markers provided for this purpose. All booklets which had page markers were examined by office staff and the score data edited where necessary. A large number of booklets were rescored by office based scorers and compared to interviewer scoring to identify any consistent errors. Any discrepancies identified in this process were resolved. Three hundred booklets were sent for international rescoring to another country. The two sets of score data were then sent independently to Statistics Canada for comparison. Any inconsistencies in scoring between countries identified through this process were examined and the source investigated. Some items were then rescored on all booklets to take account of the inconsistencies identified.

8. Weighting and grossing

The method used for estimating proficiency on the three literacy scales produces population estimates rather than individual proficiency estimates. For this reason the sample data needs to be weighted to the total population. There are two main stages in the weighting and grossing process, weighting to account for sample design and grossing to population totals at the same time as adjusting for any non-response bias.

The first stage is to apply sample weights. The sample design was such that only one person was interviewed per household. Disproportionate sampling fractions were used in both Scotland and Wales in order to ensure sufficient cases to calculate proficiency estimates for both countries. Both of these design features need to be reflected in the initial sample adjustments. When this weighting had been carried out a scaling weight was applied to reduce the number of cases to the actual achieved sample.

Distributions of the data on key characteristics were examined to identify potential sources of non-response bias, focusing in particular on characteristics thought to be associated with literacy. Factors were calculated so that the IALS proportions reflect the distributions on the Labour Force Survey for age group, sex and highest level of general education.

9. Estimating the standard errors for proficiency estimates

Estimates from sample surveys are subject to various possible sources of error. The two main types of error are systematic error and random error. The total error in a survey estimate is the difference between the estimate derived from the survey and the true value of that variable in the entire population. In undertaking the survey substantial effort is invested in guarding against systematic error, for example, ensuring interviewers administer the assessment in a standard way, ensuring no sections of the population are excluded from the sampling frame, maximising response to guard against non-response bias. Despite such precautions it is still possible that some systematic error may occur.

Sampling error is the most important component of random error. Sampling errors show the amount by which the survey estimate can be expected to differ from the true value of that variable in the population. If a survey was repeated many times under identical conditions, but using a different sample each time, then the estimates obtained from each survey would vary slightly one from the other and from the true value for the population. The amount of variation depends on the sample size and on the sample design as well as the variability of the population on the characteristic of interest. The larger the sample size the lower the sampling error is likely to be.

For estimates based on a simple random sample the calculation of the standard error is straight forward. Where the sample is stratified or the estimate is a complex statistic, the calculation of the standard error involves more complex calculation which takes into account, as in this case, the stratification and clustering of the sample design. For most surveys carried out by SSD the sampling errors are estimated by the linearisation method using a computer programme (EPSILON). The method used to produce the literacy proficiency estimates for this survey generates multiple plausible values on each of the three literacy scales for each respondent. These plausible values are used to generate literacy proficiency values for different subgroups of the population. This is a complex statistic. The Jacknife procedure is used for estimating the variance of the proficiency estimate on each scale. The sample is divided into random groups (30 in this case). Each of the random groups are in turn removed from the sample and the remaining sample reweighted to LFS estimates on the three variables, age, sex and highest level of general education. The Jacknife then derives variance estimates of the literacy proficiency estimate from each of the samples formed by excluding a random group. The full-sample estimate is calculated from the variability between the estimates.

10. Confidence intervals

Statistical theory allows us to measure the accuracy of any survey estimate. By calculating the standard error for a variable of interest, confidence intervals can be calculated around the estimate. Confidence intervals give an indication of the range in which the true population value is likely to fall. It is normal practice to refer to the 95% confidence interval which is calculated as 1.96 times the standard error and comprises the range of values 1.96 standard errors below the estimate to 1.96 standard errors above the estimate. Unlike most surveys the focus of the Adult Literacy Survey is a single area, literacy, represented by a measure on three scales. In order to remind readers of the need to take into account the error around these estimates, confidence intervals are shown for some of the key estimates in the main body of the report and standard errors are included in the tables where available.

Notes and references

1 A full description of the sample design and response rates achieved are described in the *Technical Report of the First International Adult Literacy Survey*, NCES Washington 1997.

Appendix B

Definitions and terms

Age-groups

In general the report shows data for the following age-groups:

        16-25
        26-35
        36-45
        46-55
        56-65.

In some sections age is grouped into three bands:

        16-25
        26-45
        46-65.

Benefits

People coded as being in receipt of social security benefits are those who received at least one of the following state benefits:

- unemployment benefit
- one parent benefit
- disability living allowance
- attendance allowance
- family credit
- invalid care allowance
- income support
- social fund
- incapacity benefit
- severe disablement benefit.

Document literacy (see Literacy)

Economic activity status

Economically active persons are those over the minimum school leaving age who were working or unemployed in the week before the interview. These persons constitute the labour force. People are coded as working if, in the week prior to the interview, they were:

- working either full-time or part-time;
- or on a government scheme for employment training;
- or waiting to start a new job or business;
- or they had a job or business they were away from;
- or they were doing some unpaid work either for themselves or a relative.

Using this definition students who were also working were coded as working rather than as being students.

The survey uses the ILO definition of unemployment which is a person who is out of work and has looked for work in the previous four weeks or would be available to start work in the two weeks following the interview.

People who were not working or unemployed were coded as economically inactive.

Educational attainment

Respondents were asked to give their highest educational qualification and the number of years they had spent in continuous full-time education; this information was used to derive their position on the International Standard Classification of Education (ISCED) scale. ISCED divides educational attainment into 7 categories which span three broad levels of education roughly equivalent to primary, secondary and tertiary education.

ISCED 0   Education preceding the first level, usually begins at age 3, 4 or 5 and lasts one to three years (pre-primary).
Includes those with no qualifications who finished full-time continuous education before the age of 11.

ISCED 1   First level education, usually begins at age 5, 6 or 7 and lasts for about five or six years (primary).
Includes those with no qualifications who finished full-time continuous education between the ages of 11 and 14.

ISCED 2   Second level, first stage begins at about age 11 and lasts for about 3 years (lower secondary).
Includes those whose highest qualification is one of the following:
O-level/GCSE, CSE, SCE , Intermediate GNVQ, BTEC, BEC, SCOTBEC, TEC, SCOTTEC/SCOTVEC Diploma or General Diploma or First Certificate or general certificate, all RSA qualifications, YT Certificate or any other professional, vocational or foreign qualification not coded as ISCED 3, 5, 6 or 7
or those who have completed a recognised trade apprenticeship
or those who have no qualifications but who received full-time, continuous education to

the age of 15 or 16.

ISCED 3 Second level second stage education begins at about age 14 or 15 and lasts for about three years (higher or upper secondary). Includes those whose highest qualification is one of the following:
A-level, Advanced GNVQ, BTEC, BEC, SCOTBEC, TEC, SCOTTEC/SCOTVEC National, ONC, OND, any City and Guilds qualification, Scottish Certificate of 6th year studies (Scottish CSYS) or equivalent, A/S level, or Certificate of 6th year studies (CSYS) or equivalent
or those who remained in full-time education after the age of 16.

ISCED 4 Not used

ISCED 5 Third level or higher education which leads to an award which is not equivalent to a university degree, for example a Higher National Diploma (HND).
Includes those whose highest qualification is one of the following:
Diploma in higher education, HNC/HND, Higher level of BTEC/BEC/ SCOTBEC/ TEC/ SCOTTEC/SCOTVEC, teaching qualification, nursing or other medical qualification not coded at ISCED 6 or 7, NVQ level 3 or other higher education qualifications below degree level.

ISCED 6 Third level or higher education that leads to a university degree or equivalent (includes NVQ level 4).

ISCED 7 Third level or higher education post first degree that leads to a post-graduate university degree or equivalent (includes NVQ level 5).

ISCED 9 Education not definable by level

This report uses the EUROSTAT version of the classification which means that people whose highest qualification is O-level or GCSE are coded as ISCED 2. The OECD version codes these people as ISCED 3.

The UK equivalent of ISCED is a classification based on the level of highest qualification. The use of the age at which a person first left full-time education is not a good proxy for their level of attainment because it fails to take into account education or training in later life.

Level of Highest Qualification
Higher Education & professional/vocational equivalents
Degree or equivalent
Includes: Higher degree; First degree; Professional qualifications at degree level e.g. Graduate member of professional institute; NVQ or SVQ level 4 or 5.

Other Higher Education below degree level
Includes: Diplomas in higher education and other higher education qualifications; HNC, HND Higher level BTEC; Teaching qualifications for schools or further education; Nursing, or other medical qualifications not covered above; RSA higher diploma.

A-levels, vocational level 3 & equivalents
Includes: A-level or equivalent; AS level; SCE Higher, Scottish Certificate Sixth Year Studies or equivalent; NVQ or SVQ level 3; GNVQ Advanced or GSVQ level 3; OND, ONC, BTEC National, SCOTVEC National certificate; City & Guilds advanced craft, Part III (& other names); RSA advanced diploma.

Trade Apprenticeships
GCSE/O Level grade A*-C, vocational level 2 & equivalents
Includes: NVQ or SVQ level 2, GNVQ intermediate or GSVQ level 2; RSA Diploma; City & Guilds Craft or Part II (& other names); BTEC, SCOTVEC first or general diploma etc.; O level or GCSE grade A-C, SCE Standard or Ordinary grades 1-3.

Qualifications at level 1 and below
Includes: NVQ or SVQ level 1: GNVQ Foundation level, GSVQ level 1; GCSE or O-level below grade C, SCE Standard or Ordinary below grade 3; CSE below grade 1; BTEC, SCOTVEC first or general certificate; SCOTVEC first or general certificate; SCOTVEC modules; RSA Stage 1, II or III; City & Guilds part I.

Other qualifications: level unknown
Includes: Other vocational or professional or foreign qualifications.

No qualifications

Ethnic group

Respondents were asked to say which of the following groups they considered they belonged to:

> White;
>
> Black - Caribbean;
>
> Black - African;
>
> Black - neither Caribbean or African;
>
> Indian;
>
> Pakistani;
>
> Bangladeshi;
>
> Chinese;
>
> Other.

Because of the small numbers of respondents in non-white ethnic groups the data was regrouped into just two categories: white and non-white.

Great Britain

Great Britain comprises England, Wales and Scotland.

Gross income

Gross income is personal income from all sources before deductions for income tax and National Insurance; it does not include housing benefit. In order to allow international comparisons to be made income is shown in the tables as quintiles. The quintiles represent the following gross annual income bands:

> Quintile 1 - up to £2,704pa
>
> Quintile 2 - £2,705-£5,928pa
>
> Quintile 3 - £5,929-£10,400pa
>
> Quintile 4 - £10,401-£16,848pa
>
> Quintile 5 - £16,849pa or more.

Household

A household is one person living alone or a group of people who have the address as their only or main residence and who either share one meal a day or share a living room.

Household membership

People are regarded as living at an address if they consider the address to be their main residence. There are, however, certain rules which take priority over this criterion.

- Children aged 16 or over who live away from home for the purpose of work or study and come home only for the holidays are not included at the parental address under any circumstances.

- Children of any age away from home in a temporary job and children under 16 at boarding school are always included in the parental household.
- People who have been away from the address continuously for six months or longer are excluded.
- People who have been living continuously at the address for six months or longer are included even if they have their main residence elsewhere.
- Addresses used only as second homes are never counted as main residences.

Industry

The UK Standard Industrial Classification (SIC-92) used in Chapter 3 'Literacy and work' is harmonised with both the current European Union (NACE) and International (ISIC) classifications.

United Kingdom: *Standard Industrial Classification of economic activities 1992,* Central Statistical Office 1992 (London: HMSO).

European Union: *Statistical Classification of Economic Activities* (NACE Rev.1) was published in the Official Journal of the European Communities L 293 Volume 33, 24 October 1990.

International: *The International Standard Classification of all Economic Activities* (ISIC Rev.3) was agreed by the Statistical Commission of the United Nations in 1989. All countries participating in IALS were required to provide industrial data in an earlier version of this classification (ISIC Rev. 2, 1968).

ISCED (see Educational attainment)

International Standard Classification of Occupations 1988 (ISCO 88)

ISCO is a classification of occupations developed by the International Labour Office (ILO). The 1988 version is similar but not identical in structure and detail to the Standard Occupational Classification (SOC) used in Great Britain. All countries participating in IALS were required to provide occupation data in the ISCO 88 format so as to enable international comparison (see Chapter 6). For a description of the compatibility between SOC and ISCO 88 see the *Standard Occupational Classification Volume 3,* OPCS 1991 (London : HMSO), page 25.

Literacy

The International Adult Literacy survey defines literacy as:

'Using printed and written information to function in society, to achieve one's goals, and to develop one's knowledge and potential.'

Literacy is described in terms of three domains:

Prose literacy

Prose literacy is the knowledge and skills needed to understand and use information from texts including editorials, news stories, poems and fiction;

Document literacy

Document literacy is the knowledge and skills required to locate and use information contained in various formats , including job applications, payroll forms, bus and train timetables, maps, tables and graphics;

Quantitative literacy

Quantitative literacy is the knowledge and skills required to apply arithmetic operations, either alone or sequentially, to numbers embedded in printed materials, such as balancing a cheque book, working out a tip, completing an order form or working out the amount of interest on a loan.

Occupation (see Standard Occupation Classification (SOC))

Prose literacy (see Literacy)

Qualifications (see Educational attainment)

Quantitative literacy (see Literacy)

Social class

Occupation details were collected for economically active and retired people. Unemployed people were asked about their last job and retired people were coded according to their main previous occupation. Occupations were coded according to the *Standard Occupational Classification* OPCS (HMSO, London 1991) and social class derived accordingly. As well as the individual's occupation the derivation of social class takes into account whether the person is an employee or self-employed, what his or her working status is (manager, foreman or other employee) and the size of the establishment in which he or she works.

Standard Occupational Classification (SOC)

The Standard Occupational Classification (SOC) identifies 371 different occupations in Great Britain. Jobs that require similar skills, qualifications and training have been grouped together into 9 major groups.

1. Managers and Administrators

A significant amount of knowledge and experience of the production processes, administrative procedures or service requirements associated with the efficient functioning of organisations and businesses is required for these occupations.

2. Professional Occupations

A degree or equivalent qualification is required; some occupations require post-graduate qualifications and/or a formal period of experience-related training. Examples include scientists, doctors, teachers and lawyers.

3. Associate Professional and Technical Occupations

An associated high-level vocational qualification, often involving a substantial period of full-time training or further study is required. Some additional task related training is usually provided through a formal period of induction. Examples include nurses, computer programmers and personnel officers.

4. Clerical and Secretarial Occupations

A good standard of general education is required. Certain occupations will require further additional vocational training to a well defined standard (e.g. typing or shorthand).

5. Craft and Related Occupations

A substantial period of training required, often provided by means of a work- based training programme, for example printers, carpenters and mechanics.

6. Personal and Protective Service Occupations

A good standard of general education is needed. Certain occupations will require further additional vocational training, often provided by means of a work-based training programme. Examples are police officers, chefs, hairdressers and nursery nurses.

7. Sales Occupations

A general education and a programme of work-based training related to sales procedures. Some occupations require additional specific technical knowledge but are included in this major group because the primary task involves selling

8. Plant and Machine Operatives

The knowledge and experience necessary to operate vehicles and other mobile and stationary machinery, to operate and monitor industrial plant and equipment, to assemble products from component parts according to strict rules and procedures and subject assembled parts to routine tests. Most occupations in this major group will specify a minimum standard of competence that must be attained for satisfactory performance of the associated tasks and will have an associated period of formal experience-related training.

9. Other Occupations

The knowledge and experience necessary to perform mostly simple and routine tasks involving the use of hand-held tools and in some cases, requiring a degree of physical effort. Most occupations in the major group require no formal educational qualifications but will usually have an associated short period of formal experience-related training. All non-managerial agricultural occupations are also included in this major group, primarily because of the difficulty of distinguishing between those occupations which require only a limited knowledge of agricultural techniques, animal husbandry, etc. from those which require specific training and experience in these areas. These occupations are defined in a separate minor group.

The SOC major groups were used in Chapter 3 'Literacy and work' and further detail was provided by splitting some of the major groups into smaller sub-groups where appropriate.

For a definition of the different SOC groups see the *Standard Occupational Classification Volume 1; structure of the classification,* OPCS 1990 (London : HMSO), page 7. For a description of the coding methodology see *Standard Occupational Classification Volume 3; Coding methodology,* OPCS 1991 (London : HMSO).

Unemployed persons (see Economic activity status)

United Kingdom

The United Kingdom comprises Great Britain (i.e. England, Wales and Scotland) and Northern Ireland.

Wage income

Wage income is an individual's income from employment. It did not take into account the number of hours worked. The annual wage income quintiles were calculated from the income data collected by the 1994 General Household Survey. The quintiles were:

Quintile 1 - up to £4,628pa

Quintile 2 - £4,629-£8,996pa

Quintile 3 - £8,997-£13,000pa

Quintile 4 - £13,001-£19,188pa

Quintile 5 - £19,189pa or more.

Working persons (see Economic activity status)

Appendix C
Background Questionnaire

General Information (including education)

1   ArNum   Area number.

3.   AdNum   Address number.

4.   HHNum   Household number.

5.   IntNum   Interviewer number.

6.   Time1   INTERVIEWER : PLEASE RECORD THE START TIME OF THE BACKGROUND SECTION (IN HOURS AND MINUTES - 24 HOUR CLOCK)

7.   IntDay   DATE OF INTERVIEW - DAY

8.   IntMon   DATE OF INTERVIEW - MONTH

9.   IntYr   DATE OF INTERVIEW - YEAR

10.   RelHOH   RELATIONSHIP TO HOH (CODE)

Head of Household 1
Wife of HOH 2
Cohabitee 3
Son/daughter (incl. adopted) 4
Stepson/daughter 5
Foster child 6
Son-in-law/daughter-in-law 7
Parent 8
Step-parent 9
Foster parent. 10
Parent-in-law 11
Brother/sister (incl. adopted) 12
Step-brother/sister 13
Foster brother/sister 14
Brother/sister in law 15
Grandchild 16
Grandparent 17
Other relative 18
Other non-relative 19

11.   Sex

SEX
Male 1
Female 2

12.   DOBD

DATE OF BIRTH - DAY

13.   DOBM

DATE OF BIRTH - MONTH

14.   DOBY

DATE OF BIRTH - YEAR (last 2 digits)

15.   Age

AGE ON DATE OF INTERVIEW SHOULD BE .... CHECK WITH RESPONDENT AND ENTER AGREED AGE. BUT IF DATE OF BIRTH NOT KNOWN RECORD RESPONDENT'S (OR YOUR) ESTIMATE OF AGE

16.   MarCon

MARITAL STATUS

Married 1
Living together as a couple 2

Single 3
Widowed 4
Divorced 5
Separated 6

17.   IntInstA

Now I would like to ask you a few questions about your background, your education, the languages you speak and the jobs you may have had in the past 12 months.

18.   COB

In what country were you born?

England 1
Scotland 2
Wales 3
Northern Ireland 4
Other 5

19.   COBOth   APPLIES IF Country of birth (COB) is Other

TYPE IN (MAIN) COUNTRY

20.   EdAge

How old were you when you finished your continuous full-time education? (STILL IN = 96, NEVER HAD = 97)

21   YrEd APPLIES IF Respondent received (or is receiving) a full-time education (EdAge not equal to 97)

During your lifetime, how many years of formal education have you completed beginning with the first year of primary school and not counting repeated years at the same level? (NEVER HAD AN EDUCATION = 0)

22.   QualCh APPLIES IF Respondent received (or is receiving) a full-time education (EdAge not equal to 97)

I would now like to ask you about education and work related training. Do you have..

(CODE FIRST THAT APPLIES)

Any qualifications from school or college, or connected with work or a government scheme 1
No qualifications. 2
Don't know 3

23.   QUAL APPLIES IF QualCh=1 (respondent has qualifications) or Qualch = 3 (Don't know)

SHOW CARD A. Do you have any of the qualifications shown on this card? Start at the top of the list.

(CODE UP TO 3 QUALIFICATION LEVELS FROM THE LIST)

Higher degree 1
First degree 2
Other degree level qualification such as graduate membership of professional institute 3
Diplomas in Higher Education. 4
HNC/HND, Higher level of BTEC, BEC, SCOTBEC TEC or SCOTEC/SCOTVEC 5
Teaching qualification 6
Nursing or other medical qualification not yet mentioned 7
Other Higher Education qualifications below degree level 8
RSA Higher Diploma 9
A-level or equivalent/Advanced GNVQ 10

RSA Advanced diploma/Advanced certificate 11
BTEC, BEC, SCOTBEC TEC or SCOTEC/SCOTVEC
   National/ONC/OND 12
City & Guilds advanced craft 13
Scottish Certificate of 6th Year Studies (Scottish
   CSYS) or equivalent. 14
SCE(Higher) or equivalent 15
A/S level/ Certificate of 6th Year Studies(CSYS)
   or equivalent. 16
RSA diploma 17
City & Guilds craft 18
BTEC, BEC, SCOTBEC TEC or SCOTVEC First
   diploma or General diploma 19
O-level/GCSE grades A B C/ SCE Standard grades
   1 2 3/SCE Ordinary grades A B C/CSE grade 1
   or equivalent CSE/GCSE/SCE/Intermediate GNVQ .. 20
CSE, GCSE,SCE not yet mentioned............................. 21
BTEC, BEC, SCOTBEC TEC or SCOTEC/SCOTVEC
   First certificate or general certificate 22
YT certificate 23
SCOTVEC National certificate modules......................... 24
RSA other qualifications (including Stage I, II, III) 25
City & Guilds other qualification 26
Any other professional/ vocational qualification/foreign
   qualifications 27
None of these/foundation GNVQ 28
Don't know   29

24.   OLevel   APPLIES IF respondent has O-level(s) or
                       equivalent (Qual=20)

             SHOW CARD B.

             Which examinations on this card have you passed?
             CODE ALL THAT APPLY.

             CSE grade 1 1
             GCSE - grade A B C.. 2
             GCSE - grade D E F G 3
             GCE O level - obtained before 1975 4
             GCE O level - obtained 1975 or later - grades A B C 5
             GCE O level - obtained 1975 or later - grades D E 6
             SCE Ordinary - obtained before 1973 7
             SCE Ordinary - obtained 1973 or later - Bands A B C 8
             SCE Ordinary - obtained 1973 or later - Bands D E 9
             SCE Standard Grade, level 1-3..................................... 10
             SCE Standard Grade, level 4-7..................................... 11

25.   OLEV1   APPLIES IF Olevel=1

             How many subjects at CSE grade 1 did you
             pass in?

26.   OLEV2   APPLIES IF Olevel=2

             How many subjects at GCSE grade A B C did
             you pass in?

27.   OLEV3   APPLIES IF Olevel=3

             How many subjects at GCSE grade D E F G
             did you pass in?

28.   OLEV4   APPLIES IF Olevel=4

             How many subjects at GCE O level (obtained
             before 1973) did you pass in?

29.   OLEV5   APPLIES IF Olevel=5

             How many subjects at GCE O level grade A B C
             did you pass in?

30.   OLEV6   APPLIES IF Olevel=6

             How many subjects at GCE O level grade D E
             did you pass in?

31.   OLEV7   APPLIES IF Olevel=7

             How many subjects at SCE Ordinary (obtained
             before 1973) did you pass in?

32.   OLEV8   APPLIES IF Olevel=8

             How many subjects at SCE Ordinary Bands A B
             C did you pass in?

33.   OLEV9   APPLIES IF Olevel=9

             How many subjects at SCE Ordinary Bands D
             E did you pass in?

34.   OLEV10   APPLIES IF Olevel=10

             How many subjects at SCE Standard Grades
             1-3 did you pass in?

35.   OLEV11   APPLIES IF Olevel=11

             How many subjects at SCE Standard Grades
             4-7 did you pass in?

36.   NVQSVQ   APPLIES IF QualCh=1 (respondent has
                      qualifications) or Qualch=3 (Don't know)

             (A new system of national vocational qualifications has
             recently been introduced called NVQs and, in Scotland,
             SVQs.) Do you have any FULL NVQs or FULL SVQs?

             Yes. 1
             No.. 2
             Don't know 3
             Never heard of NVQs/SVQs. 4

37.   NVQlev   APPLIES IF respondent has full NVQs or SVQs
                      (NVQSVQ=1)

             What is your highest level of full NVQ/SVQ?

             level 1 1
             level 2 2
             level 3 3
             level 4 4
             level 5 5
             Don't know 6

38   SCHOOL   APPLIES IF respondent is still in full-time
                      education (EdAge=96)

             (Ask or record 'Other' if person aged 20 years or more)

             Are you still at school or are you in some kind of full-time
             education?

             School 1
             Other full-time education 2

39.   QAPPREN   APPLIES IF respondent finished full-time
                      education between ages of 5 and 65
                      (EdAge=5-65) or never had full-time education
                      (EdAge=97) or in full-time education but not at
                      school (School=2).

             Are you doing or have you completed, a recognised trade
             apprenticeship?

             Yes (completed) 1
             Yes (still doing) 2
             No (including apprenticeship begun but discontinued) 3

40.   WhyStop   APPLIES IF the highest education level the
                      respondent finally completed was one of the
                      following (international education levels
                      (ISCED) 0,1,2) OR respondent in full-time
                      education but not at school (School=2).

ISCED 0 - respondent has no qualifications (QualCh=2) and left school before age of 11 (EdAge<11).

ISCED 1 - a) respondent has no qualifications (Qualch=2) and left school between ages of 11 and 14 (EdAge=11-14) OR b) respondent has no qualifications (QualCh=2) and still in school (EdAge=96).

ISCED 2 - a) Respondent has one of the following qualifications; RSA diplomas or certificates, BTEC diploma or equivalent, O level or equivalent, CSE, BTEC certificate or diploma, YT certificate or other professional/vocational qualifications (Qual=9,11,17,19,20,21,22,23,25,27); or has completed recognised trade apprenticeship (QAPPREN=1) OR b) Respondent has no qualifications (QualCh=2) and left school at age of 15 or 16 (EdAge=15,16).

What was the main reason you stopped your schooling when you did?

School or further education not available / not accessible	1
Had enough education	2
Had to work / financial reasons	3
Wanted to work / wanted to learn a trade	4
Family reasons, help family business, illness at home, marriage, pregnancy etc	5
Did not like school	6
Did not do well in school / boredom	7
Personal illness / disability	8
To join the military	9
Don't know	10
Other - please specify at next question	11

41.   WhyStopO   APPLIES IF WhyStop=Other

What was this reason?

Language Information

42.   WhatLa

What language did you FIRST speak as a child?
LIST IS IN ALPHABETICAL ORDER.
ACCEPT TWO RESPONSES ONLY IF LANGUAGES WERE SPOKEN EQUALLY. CODE UP TO TWO LANGUAGES. (84 = OTHER)

English	1	Irish	31
Welsh	2	Italian	32
Afrikaans	3	Japanese	33
Albanian	4	Kurdic	34
Arabic	5	Lebanese	35
Awadhi	6	Lingala	36
Belorussian	7	Malay	37
Bengali	8	Mandarin	38
Bihari	9	Marathi	39
Bulgarian	10	Mende	40
Cantonese	11	Menon	41
Chinese–other	12	Metis	42
Czech	13	Mossi	43
Danish	14	Moroccan	44
Dievehi	15	Norwegian	45
Dutch	16	Pahari	46
Egyptian	17	Persian	47
Finnish	18	Polish	48
Flemish	19	Portuguese	49
French	20	Punjabi	50
Fulani	21	Pushto	51
Ganda	22	Ragestani	52
German	23	Romansh	53
Greek	24	Romanian	54
Gujerati	25	Russian	55
Hebrew	26	Rwanda	56
Hindi	27	Sango	57
Hungarian	28	Scots–gallic	58
Indonesian	29	Serbo-Croat	59
Iranian	30	Shona	60

Sindi	61	Turkish	74
Shingalese	62	Ukranian	75
Slovak	63	Urdu	76
Slovenian	64	Vietnamese	77
Soleti	65	Wolof	78
Somali	66	Wu	79
Sorbian	67	Xhose	80
Spanish	68	Yiddish	81
Swahili	69	Yoruba	82
Swedish	70	Zulu	83
Swiss-German	71	Other - specify at next question	84
Tamil	72		
Telugu	73		

43.   LangOth   APPLIES IF respondent's first language not on list above (WhatLa=84).

What was this other language?

44.   LangMore   APPLIES IF respondent's first language not on list above (WhatLa=84).

INTERVIEWER ANY OTHER LANGUAGES?

Yes	1
No	2

45.   LangOth2   APPLIES IF respondent has other first languages to record (LangMore=1).

What was this other language?

46.   SpeaLn1   APPLIES IF one of respondent's first languages spoken as a child was Welsh (WhatLa=2)

How would you rate your current ability to speak Welsh?

Cannot speak Welsh	1
Poor	2
Fair	3
Good	4
Very good	5

47.   UndLn1   APPLIES IF one of respondent's first languages spoken as a child was Welsh (WhatLa=2)

How would you rate your current ability to understand Welsh when it is spoken to you?

Cannot understand Welsh	1
Poor	2
Fair	3
Good	4
Very good	5

48.   ReadLn1   APPLIES IF one of respondent's first languages spoken as a child was Welsh (WhatLa=2)

How would you rate your current reading skills in Welsh?

Cannot read in Welsh	1
Poor	2
Fair	3
Good	4
Very good	5

49.   WritLn1   APPLIES IF one of respondent's first languages spoken as a child was Welsh (WhatLa=2)

How would you rate your current writing skills in Welsh?

Cannot write in Welsh	1
Poor	2
Fair	3
Good	4
Very good	5

50. NoEng   APPLIES IF respondent only spoke first language(s) other than English as a child (WhatLa not equal to 1).

INTERVIEWER: DOES RESPONDENT SPEAK ANY ENGLISH?

Yes 1
No 2

51. AgeEng   APPLIES IF respondent speaks English now but this was not one of the first languages spoken as a child (NoEng=1).

How old were you when you first started to learn English?

INTERVIEWER RECORD YOUNGEST AGE MENTIONED

52. Convers

What languages (including English) do you speak well enough to conduct a conversation?

CODE UP TO SIX LANGUAGES. (84 = OTHER)

List of languages the same as Q.42 WhatLa

53. OthCon   APPLIES IF respondent can conduct a conversation with language not on list above (Convers=84).

What was this other language?

54. LangMor4   APPLIES IF respondent can conduct a conversation with language not on list above (Convers=84).

INTERVIEWER ANY OTHER LANGUAGES?

Yes 1
No 2

55. OthCon2   APPLIES IF LangMor4=Yes

What was this other language?

56. LangHome   APPLIES IF respondent can conduct a conversation in more than one language, including Welsh (Convers=2)

What language do you speak most often at home?

(List of up to 6 languages mentioned at Question Convers)

(language 1) 1
(language 2) 2
(language 3) 3
(language 4) 4
(language 5) 5
(language 6) 6

57. LangWork APPLIES IF respondent can conduct a conversation in more than one language, including Welsh (Convers=2)

What language do you speak most often at work or school?

(List of up to 6 languages mentioned at Question Convers)

(language 1) 1
(language 2) 2
(language 3) 3
(language 4) 4
(language 5) 5
(language 6) 6
Not Applicable (neither work nor go to school) 7

58. LangLeis   APPLIES IF respondent can conduct a conversation in more than one language, including Welsh (Convers=2)

What language do you speak most often during leisure activities?

(List of up to 6 languages mentioned at Question Convers)

(language 1) 1
(language 2) 2
(language 3) 3
(language 4) 4
(language 5) 5
(language 6) 6

59. LangExpr   APPLIES IF respondent can conduct a conversation in more than one language, including Welsh (Convers=2)

In what language can you express yourself most easily?

(List of up to 6 languages mentioned at Question Convers)

(language 1) 1
(language 2) 2
(language 3) 3
(language 4) 4
(language 5) 5
(language 6) 6

Labour Force Information

60. Curstat

I would now like to talk to you about your employment status. Did you do any paid work in the 7 days ending last Sunday either as an employee or as self-employed?

Yes 1
No 2

61. Govsch   APPLIES IF respondent did not do any paid work in last week (Curstat=2).

Were you on a government scheme for employment training?

Yes 1
No 2

62. Busaway   APPLIES IF respondent not on a government training scheme (Govsch=2)

Did you have a job or business that you were away from?

Yes 1
No 2
Waiting to take up new job/business 3

63. Unpaid   APPLIES IF respondent not away from job or business (Busaway=2)

Did you do any unpaid work in that week for any business that you or a relative owns?

Yes, a business that you own 1
Yes, a business that a relative owns 2
No 3

64. Lookwk   APPLIES IF respondent has not worked in last week (Curstat=2), AND not on government training scheme (Govsch=2), AND not away from job or business (Busaway=2), AND not doing any unpaid work (Unpaid=3).

Thinking of the 4 weeks ending last Sunday, were you looking for any kind of paid work or government training scheme at any time in those 4 weeks?

Yes 1
No 2

65. Govaval APPLIES IF respondent looking for work (Lookwk=1).

If a job or a place on a government scheme had been available in the week ending last Sunday would you have been available to start within two weeks?

Yes 1
No 2

66. Econact APPLIES IF respondent not looking for work (Lookwk=2) OR looking for work but not available to start within 2 weeks (Govaval=2).

What was the main reason you did not seek any work in the last 4 weeks (would not be able to start within two weeks)? SHOW CARD C

Student 1
Waiting to take up a job or business 2
Temporarily sick or injured 3
Long term sick or disabled 4
Child care responsibilities 5
Other personal or family responsibilities 6
Looking after the family/home 7
Retired from paid work. 8
Not interested in working 9
Other reasons 10

67. Work12m APPLIES IF respondent has not worked in last week (Curstat=2) AND not away from job or business(Busaway=2) AND not on government training scheme (Govsch=3).

Did you work at a job or business at any time in the past 12 months (regardless of the number of hours per week)?

Yes 1
No 2

68. EverWk APPLIES IF respondent has not worked at any time in past 12 months (Work12m=2).

Have you ever had a paid job, or place on a government scheme apart from casual or holiday work (or the job that you were waiting to take up)?

Yes 1
No 2

69. LeftYr APPLIES IF respondent has had a job (EverWk=1).

Which year did you leave your last PAID job or last have a business?

70. NumJob APPLIES IF respondent is currently employed (Curstat=1) OR has worked in last 12 months (Work12m=1).

How many different employers have you had in the past 12 months?

71. WorkFTPT APPLIES IF respondent is currently employed (Curstat=1) OR has worked in last 12 months (Work12m=1).

In your (main) job were you working:

full time 1
or part time? 2

72. YPtJob APPLIES IF respondent working part time (WorkFTPT=2)

Why did you work part time?

Own illness or disability 1
Child care responsibilities 2

Other family responsibilities 3
Going to school/ taking training 4
Could only find a part-time job 5
Did not want to work full time 6
Other - please specify at next question 7

73. YPtOth APPLIES IF 'other' reason for working part time (YPtJob=7)

What was the reason for working part time?

74. IntInst APPLIES IF respondent currently employed (Curstat=1) OR worked in last 12 months (Work12m=1) OR has ever worked (Everwk=1).

I would now like to ask you some questions about your(main) job, (that is the job you worked the most hours for).

75. IndD APPLIES IF respondent currently employed (Curstat=1) OR worked in last 12 months (Work12m=1) OR has ever worked (Everwk=1).

What did the firm/organisation you worked for mainly make or do (at the place where you worked)?

DESCRIBE FULLY - PROBE MANUFACTURING or PROCESSING or DISTRIBUTION ETC. AND MAIN GOODS PRODUCED, MATERIALS USED, WHOLESALE or RETAIL ETC

76. IndT APPLIES IF respondent currently employed (Curstat=1) OR worked in last 12 months (Work12m=1) OR has ever worked (Everwk=1).

ENTER A TITLE FOR THE INDUSTRY

77. OccT APPLIES IF respondent currently employed (Curstat=1) OR worked in last 12 months (Work12m=1) OR has ever worked (Everwk=1).

What was your (main) job?

ENTER JOB TITLE

78. OccD APPLIES IF respondent currently employed (Curstat=1) OR worked in last 12 months (Work12m=1) OR has ever worked (Everwk=1).

What did you mainly do in your job?

CHECK SPECIAL QUALIFICATIONS/TRAINING NEEDED TO DO THE JOB

79. Stat APPLIES IF respondent currently employed (Curstat=1) OR worked in last 12 months (Work12m=1) OR has ever worked (Everwk=1).

Were you working as an employee or were you self-employed?

Employee 1
Self-employed 2

80. Manage APPLIES IF respondent is an employee (Stat=1)

ASK OR RECORD
Did you have any managerial duties, or were you supervising any other employees?

Manager 1
Foreman/supervisor 2
Not manager/supervisor 3

81. ManagNo APPLIES IF respondent was manager or supervisor (Manage=1,2)

How many employees did you have supervisory or managerial responsibility for?

5 or more 1
Less than 5. 2

82. Empno   APPLIES IF respondent is an employee (Stat=1)

How many employees were there at the place where you worked?

1-24 1
25 or over 2

83. EmpBrit   APPLIES IF respondent is an employee (Stat=1)

In total, about how many persons are employed by this business at all locations in Britain?

Less than 20 1
20 to 99 2
100 to 199 3
200 to 499 4
500 or over 5

84. Solo   APPLIES IF respondent is self-employed (Stat=2)

ASK OR RECORD
Were you working on your own or did you have employees?

On own/with partner(s) but no employees 1
With employees 2

85. SENo   APPLIES IF respondent is self-employed with employees (Solo=2)

How many people did you employ at the place where you worked?

1-24 1
25 or over 2

86. JobTyp   APPLIES IF respondent is employed (Curstat=1) OR has worked in last 12 months (Work12m=1)

Leaving aside your own personal intentions and circum-stances, was your job:

a permanent job or work contract of unlimited duration ... 1
or was there some way that it was NOT permanent? 2

87. TotHrUs   APPLIES IF respondent is employed (Curstat=1) OR has worked in last 12 months (Work12m=1)

How many hours per week do/did you usually work in your (main) job/business - please exclude meal breaks and overtime?

97 HOURS OR MORE = 97
DON'T KNOW OR REFUSAL = 99

88. WeekWk   APPLIES IF respondent is employed (Curstat =1) OR has worked in last 12 months (Work12m=1)

During the past 12 months, how many weeks did you work (at all jobs including time off for holidays, maternity leave, illness and industrial action)?

89. WantWk   APPLIES IF respondent has worked less than 52 weeks in last 12 months (WeekWk<52) OR respondent has ever had job (Everwk=1)

During the past 12 months, in the weeks when you were without work, did you want to work?

Yes 1
No 2

90. NoWant   APPLIES IF respondent did not want to work (WantWk=2)

May I just check, what was the main reason that you did not want to work?

Own illness or disability 1
Childcare responsibilities 2
Other personal or family responsibilities 3
Going to school or taking training 4
Retired from paid work 5
Not interested in working 6
Homemaker 7
Any other reason- please specify at next question 8

91. YNotOth   APPLIES IF respondent had 'other' reason for not working (NoWant=8)

What was the reason for not wanting to work?

92. Lookwork   APPLIES IF respondent wanted to work (WantWk=1).

During the last 12 months, for how many weeks were you without work and NOT looking for work?

93. NoLook   APPLIES IF respondent not looking for work during any weeks of last 12 months (LookWork>0).

May I just check, what was the main reason that you did not look for work during these weeks?

Own illness or disability 1
Childcare responsibilities 2
Other personal or family responsibilities 3
Waiting for the results of an application for a job/ being assessed by an ET training agent 4
Waiting for a job to start 5
Did not have the skills or experience for available jobs ... 6
Going to school or taking training 7
Too old to work/Retired from paid work 8
Not interested in working 9
Any other reason- please specify at next question 10

94. YLookO   APPLIES IF respondent had 'other' reason for not working (NoLook=10)

What was the reason for not looking for work?

Reading and writing at work

95. ULetter   APPLIES IF respondent currently employed (Curstat=1) OR working on government training scheme (Govsch=1) OR away from job or business (Busaway=1) OR worked in last 12 months (work12m=1).

The following questions refer to the job at which you worked the most hours in the last 12 months.

How often (do/did) you read or use information from each of the following as part of your main job?

Letters or memos? SHOW CARD D
Every day 1
A few times a week 2
Once a week 3
Less than once a week 4
Rarely or never 5

96. Report — APPLIES IF respondent currently employed (Curstat=1) OR working on government training scheme (Govsch=1) OR away from job or business (Busaway=1) OR worked in last 12 months (work12m=1).

Reports, articles, magazines or journals? SHOW CARD D
Every day 1
A few times a week 2
Once a week 3
Less than once a week 4
Rarely or never 5

97. Manual — APPLIES IF respondent currently employed (Curstat=1) OR working on government training scheme (Govsch=1) OR away from job or business (Busaway=1) OR worked in last 12 months (work12m=1).

Manuals or reference books, including catalogues?
SHOW CARD D

Every day 1
A few times a week 2
Once a week 3
Less than once a week 4
Rarely or never 5

98. Diagram — APPLIES IF respondent currently employed (Curstat=1) OR working on government training scheme (Govsch=1) OR away from job or business (Busaway=1) OR worked in last 12 months (work12m=1).

Diagrams? SHOW CARD D

Every day 1
A few times a week 2
Once a week 3
Less than once a week 4
Rarely or never 5

99. Bills — APPLIES IF respondent currently employed (Curstat=1) OR working on government training scheme (Govsch=1) OR away from job or business (Busaway=1) OR worked in last 12 months (work12m=1).

Bills, invoices, spreadsheets or budget tables?
SHOW CARD D

Every day 1
A few times a week 2
Once a week 3
Less than once a week 4
Rarely or never 5

100. Material — APPLIES IF respondent currently employed (Curstat=1) OR working on government training scheme (Govsch=1) OR away from job or business (Busaway=1) OR worked in last 12 months (work12m=1).

Material written in a language other than English?
SHOW CARD D

Every day 1
A few times a week 2
Once a week 3
Less than once a week 4
Rarely or never 5

101. Direct — APPLIES IF respondent currently employed (Curstat=1) OR working on government training scheme (Govsch=1) OR away from job or business (Busaway=1) OR worked in last 12 months (work12m=1).

Directions or instructions for medicines, recipes, or other products? SHOW CARD D
Every day 1
A few times a week 2
Once a week 3
Less than once a week 4
Rarely or never 5

102. EComput — APPLIES IF respondent currently employed (Curstat=1) OR working on government training scheme (Govsch=1) OR away from job or business (Busaway=1) OR worked in last 12 months (work12m=1).

Read or use information from computers? SHOW CARD D

Every day 1
A few times a week 2
Once a week 3
Less than once a week 4
Rarely or never 5

103. WLetter — APPLIES IF respondent currently employed (Curstat=1) OR working on government training scheme (Govsch=1) OR away from job or business (Busaway=1) OR worked in last 12 months (work12m=1).

How often (do/did) you write or fill out each of the following as part of your main job?

Letters or memos? SHOW CARD D

Every day 1
A few times a week 2
Once a week 3
Less than once a week 4
Rarely or never 5

104. Forms — APPLIES IF respondent currently employed (Curstat=1) OR working on government training scheme (Govsch=1) OR away from job or business (Busaway=1) OR worked in last 12 months (work12m=1).

Forms or things such as bills, invoices or budgets? SHOW CARD D

Every day 1
A few times a week 2
Once a week 3
Less than once a week 4
Rarely or never 5

105. Article — APPLIES IF respondent currently employed (Curstat=1) OR working on government training scheme (Govsch=1) OR away from job or business (Busaway=1) OR worked in last 12 months (work12m=1).

Reports or articles? SHOW CARD D

Every day 1
A few times a week 2
Once a week 3
Less than once a week 4
Rarely or never 5

106. Estimate   APPLIES IF respondent currently employed (Curstat=1) OR working on government training scheme (Govsch=1) OR away from job or business (Busaway=1) OR worked in last 12 months (work12m=1).

Estimates or technical specifications? SHOW CARD D

Every day 1
A few times a week 2
Once a week 3
Less than once a week 4
Rarely or never 5

107. Size   APPLIES IF respondent currently employed (Curstat=1) OR working on government training scheme (Govsch=1) OR away from job or business (Busaway=1) OR worked in last 12 months (work12m=1).

In your main job, how often do you use arithmetic or mathematics (that is, adding, subtracting, multiplying or dividing) to:
measure or estimate the size or weight of objects?
SHOW CARD D

Every day 1
A few times a week 2
Once a week 3
Less than once a week 4
Rarely or never 5

108. Price   APPLIES IF respondent currently employed (Curstat=1) OR working on government training scheme (Govsch=1) OR away from job or business (Busaway=1) OR worked in last 12 months (work12m=1).

calculate prices, costs or budgets? SHOW CARD D

Every day 1
A few times a week 2
Once a week 3
Less than once a week 4
Rarely or never 5

109. ReadMJ   APPLIES IF respondent currently employed (Curstat=1) OR working on government training scheme (Govsch=1) OR away from job or business (Busaway=1) OR worked in last 12 months (work12m=1).

How would you rate your reading skills in English for your main job? Would you say they were

Excellent 1
Good 2
Moderate 3
Poor? 4
No opinion/not applicable (DO NOT PROMPT) 5

110. ReadJOpp   APPLIES IF respondent currently employed (Curstat=1) OR working on government training scheme (Govsch=1) OR away from job or business (Busaway=1) OR worked in last 12 months (work12m=1).

To what extent are your reading skills in English limiting your job opportunities - for example, advancement or getting another job? Are they...

greatly limiting 1
somewhat limiting 2
not at all limiting? 3

111. WriteMJ   APPLIES IF respondent currently employed (Curstat=1) OR working on government training scheme (Govsch=1) OR away from job or business (Busaway=1) OR worked in last 12 months (work12m=1).

How would you rate your writing skills in English for your main job? Would you say they are...

excellent 1
good 2
moderate 3
poor? 4
No opinion/not applicable (DO NOT PROMPT) 5

112. WritJOpp   APPLIES IF respondent currently employed (Curstat=1) OR working on government training scheme (Govsch=1) OR away from job or business (Busaway=1) OR worked in last 12 months (work12m=1).

To what extent are your writing skills in English limiting your job opportunities - for example, advancement or getting another job? Are they...

greatly limiting 1
somewhat limiting 2
not at all limiting? 3

113. Math   APPLIES IF respondent currently employed (Curstat=1) OR working on government training scheme (Govsch=1) OR away from job or business (Busaway=1) OR worked in last 12 months (work12m=1).

How would you rate your mathematical skills for your main job? Would you say they are..

excellent 1
good 2
moderate 3
poor? 4
No opinion/not applicable (DO NOT PROMPT) 5

114. MathJOpp   APPLIES IF respondent currently employed (Curstat=1) OR working on government training scheme (Govsch=1) OR away from job or business (Busaway=1) OR worked in last 12 months (work12m=1).

To what extent are your mathematical skills limiting your job opportunities - for example, advancement or getting another job? Are they...

greatly limiting 1
somewhat limiting 2
not at all limiting? 3

Adult education and training

115. AdultInt

The following questions will deal with any education or training which you may have taken in the past 12 months.

During the last 12 months did you receive any training or education including courses, private lessons, correspondence courses, workshops, on-the-job training apprenticeship training, arts, crafts, recreation courses or any other training or education? SHOW CARD E.

Yes 1
No 2

116. TotCours   APPLIES IF respondent received any adult education or training in last 12 months (AdultInt=1)

In total, how many courses did you take in the past 12 months?

117. WhatC1   APPLIES IF respondent received any adult education or training in last 12 months (AdultInt=1)

SHOW CARD F. Here is a list of types of course. Which best describes the first of these courses?

(NOTE, IF THE RESPONDENT LISTS MANY COURSES THAT ALL BEGIN AT THE SAME TIME, CHECK WHETHER THEY ARE UNITS OF ONE AND ENTER OVERALL COURSE TITLE)

Skill training/upgrading related to your
    profession or occupation 1
Skill training/upgrading not related to your
    occupation or profession 2
Personal development and communication
    skills (e.g. time management, team leadership,
    stress management) 3
Management/organisation training and
    development (include Human resource management) . 4
Computer software training (e.g. Wordperfect,
    Excel, pagemaker) 5
Health and safety (e.g. first aid, lifting and
    handling goods, risk assessment) 6
Languages (business and conversational) 7
Operation and/or maintenance of equipment 8
Quality assurance/control 9
Sport and physical fitness 10
Other recreational activities (e.g. Bridge or painting) 11
Preliminary education and training 12
None of these 13

118. MFS   APPLIES IF course taken was skill training or academic subject (WhatC=1,2,12)

SHOW CARD G. Which of these subject areas was this training or education?

Education, recreational and counselling services 1
Fine and applied arts 2
Humanities and related fields 3
Social sciences and related fields 4
Commerce, management and business administration ... 5
Agricultural and biological sciences and technologies 6
Engineering and applied sciences 7
Engineering and applied sciences,
    technologies and trades 8
Health professions, sciences and technologies 9
Mathematics and physical sciences. 10
No specialisation 11
Other 12

119. CMoney   APPLIES IF respondent received any adult education or training in last 12 months (AdultInt=1)

Now I would like to ask you about the (1st course)
Was this training or education financially supported
by...(SHOW CARD H - CODE ALL THAT APPLY)

An employer, or potential employer 1
Training for Work 2
Other government or local authority organisation 3
Yourself, or family, or relative 4
A union or professional organisation 5
Other (NO PROMPT) 6
No fees (NO PROMPT) 7
Don't know (NO PROMPT) 8

120. CMGovt   APPLIES IF course is supported by 'Other government or local authority organisation'(Cmoney=3)

Which government or local authority organisation supported this course? SPECIFY

121. CMOth   APPLIES IF course is supported by 'Other'(CMoney=6)

CODED OTHER AT CMONEY - SPECIFY

122. CQuali   APPLIES IF respondent received any adult education or training in last 12 months (AdultInt=1)

Was/Is this education or training towards qualifications?

Yes 1
No 2

123. TAPPREN   APPLIES IF respondent received any adult education or training in last 12 months (AdultInt=1)

ASK OR RECORD. Was the education or training towards a recognised trade apprenticeship?

Yes 1
No 2

124. Prfss   APPLIES IF respondent's education or training was not towards qualifications (CQuali=2).

Was/Is this education or training towards professional or career upgrading?

Yes 1
No 2

125. WhQual   APPLIES IF respondent's education or training was towards qualifications (CQuali=1).

What qualifications does/did this course lead to?
SHOW CARD I (CODE FIRST THAT APPLIES)

Higher degree 1
First degree 2
Other degree level qualification such as graduate
    membership of professional institute 3
Diplomas in Higher Education 4
HNC/HND, Higher level of BTEC, BEC, SCOTBEC,
    TEC or SCOTTEC/SCOTVEC 5
Teaching qualifications 6
Nursing or other medical qualifications not yet
    mentioned 7
Other Higher Education qualifications below degree
    level 8
RSA higher diploma 9
A-level or equivalent/Advanced GNVQ 10
RSA Advanced diploma/Advanced certificate 11
BTEC, BEC, SCOTBEC TEC or SCOTEC/SCOTVEC
    National/ONC/OND 12
City and Guilds advanced craft 13
Scottish Certificate of 6th Year studies Scottish
    CSYS or equivalent 14
SCE Higher or equivalent 15
A/S level/Certificate of 6th Year studies CSYS or
    equivalent 16
RSA diploma 17
City & Guilds craft 18
BTEC, BEC, SCOTBEC TEC or
    SCOTEC/SCOTVEC First Diploma or
    General Diploma 19
O-level/ GCSE grades A B C/SCE Standard
    grades 1 2 3/SCE Ordinary grades A B C/CSE grade 1
    or equivalent CSE/GCSE/SCE/Intermediate GNVQ .. 20
CSE,GCSE,SCE not yet mentioned 21
BTEC, BEC, SCOTBEC TEC or SCOTEC/SCOTVEC
    First certificate or General certificate 22
YT certificate 23

SCOTVEC National Certificate modules 24
RSA other qualification including Stage I, II, III 25
City & Guilds other qualification 26
Any other professional/vocational
    qualification/foreign qualifications 27
Foundation GNVQ/none of these 28
Don't know 29

126. CByW   APPLIES IF respondent received any
                     adult education or training in last 12 months
                     (AdultInt=1)

Was this training or education given by
(SHOW CARD J - CODE ALL THAT APPLY)

A University (including Open University) or other
    higher education establishment 1
A further education college or other adult
    education college/centre... 2
A commercial organisation (for example, a private
    training provider) 3
A producer or supplier of equipment 4
A non-profit organisation such as an employer
    association, voluntary organisation or a trade union 5
An employer or a parent company 6
Other government or local authority organisation 7
Other provider (NO PROMPT) 8

127. CWGovt   APPLIES IF respondent's course was given by
                     an 'other government or local authority
                     organisation' (CByW=7).

Please specify which other government or local authority
organisation gave this training?

128. CCCWOth   APPLIES IF respondent's course was given by
                     'Other provider' (CByW=8).

CODED OTHER AT CByW SPECIFY?

129. CWhere   APPLIES IF respondent received any adult
                     education or training in last 12 months
                     (AdultInt=1)

Where did this training or education take place?
(CODE ONE ONLY)

University or other higher education establishment 1
Further Education college or other adult education
    college/centre 2
Primary or secondary school... 3
Business or Commercial school 4
Work 5
Training Centre 6
Conference centre or hotel... 7
Community Centre or sports facility 8
Home (Open University or other
    correspondence course) 9
Employment rehabilitation centre/community project 10
Elsewhere 11

130. NWkT   APPLIES IF respondent received any adult
                     education or training in last 12 months
                     (AdultInt=1)

For how many weeks did this training or education last?
(ENTER 1 IF LESS THAN A WEEK)

131. NDaysT   APPLIES IF respondent received any adult
                     education or training in last 12 months
                     (AdultInt=1)

On average, how many days per week was it?
(ENTER 1 IF LESS THAN A DAY)

132. NHrT   APPLIES IF respondent received any adult
                     education or training in last 12 months
                     (AdultInt=1)

On average, how many hours per day was it?
(ENTER 1 IF LESS THAN 1 HOUR)

133. MReasTEd APPLIES IF respondent received any adult
                     education or training in last 12 months
                     (AdultInt=1)

What was the main reason you took this training or
education? Was it for... RUNNING PROMPT

career/job related purposes 1
personal interest? 2
Other (NO PROMPT) 3

134. TWrk   APPLIES IF course taken for career/job related
                     purposes (MReasTEd=1)

To what extent are you using the skills or knowledge
acquired in this training at work? (READ CATEGORIES)

To a great extent 1
Somewhat 2
Very little 3
Not at all? 4
Not applicable (NO PROMPT) 5

135. SuggT   APPLIES IF respondent received any adult
                     education or training in last 12 months
                     (AdultInt=1)

Who suggested you take this training or education?

CODE ALL THAT APPLY

You did 1
Your friends or family 2
Your employer 3
Other employees 4
Part of a collective agreement 5
Your union or trade association 6
Legal or professional requirement 7
Social services or Job centre 8
Other (NO PROMPT) 9
Don't know (NO PROMPT) 10

136. ProvT   APPLIES IF respondent received any adult
                     education or training in last 12 months
                     (AdultInt=1)

Was this training or education provided through.

SHOW CARD K - CODE ALL THAT APPLY

Classroom instruction, seminars or workshops 1
Educational software 2
Radio or TV broadcasting 3
Audio/video cassettes, tapes or disks 4
Reading materials 5
On-the-job training 6
Other methods (NO PROMPT) 7

137. WhatC2   APPLIES IF respondent has taken more than
                     one course in last 12 months (TotCours>=2)

SHOW CARD F. Here is a list of types of courses. Which best
describes the 2nd of these courses?

138. WhatC3   APPLIES IF respondent has taken more than
                     two courses in last 12 months (TotCours>=3)

SHOW CARD F. Here is a list of types of courses. Which best
describes the 3rd of these courses?

REPEAT QUESTIONS MFS TO ProvT FOR UP TO 3 COURSES

139. EdWant

In the last 12 months was there any training course or education that you WANTED to take for career or job-related reasons but did not?

Yes 1
No 2

140. RNoEd   APPLIES IF respondent wanted to take course for career related purposes but did not (EdWant=1)

SHOW CARD L. What were the reasons you did not take this training or education?

CODE ALL THAT APPLY

Too busy/lack of time 1
Too busy at work 2
Course not offered 3
Family responsibilities 4
Too expensive/no money 5
Lack of qualifications 6
Lack of employer support 7
Course offered at inconvenient time 8
Language reasons 9
Health reasons 10
Other (specify at next question) 11

141. HobWant

In the last 12 months was there any other training course that you WANTED to take but did not, such as hobby, recreational or interest courses?

Yes 1
No 2

142. RNoHob   APPLIES IF respondent wanted to take recreational course but did not (HobWant=1)

SHOW CARD L. What were the reasons you did not take this training or education?

CODE ALL THAT APPLY

Too busy/lack of time 1
Too busy at work 2
Course not offered 3
Family responsibilities 4
Too expensive/no money 5
Lack of qualifications 6
Lack of employer support 7
Course offered at inconvenient time 8
Language reasons 9
Health reasons 10
Other 11

Reading and writing at home

143. Lib

The next few questions deal with reading and writing in your daily life excluding work (or school).
I am going to read you a list of activities.
Please tell me if you do each of them daily, weekly, every month, several times a year or never?

How often do you...

Use a public library? SHOW CARD M

Daily 1
Weekly 2
Monthly 3
Several times a year 4
Never 5

144. Film

Attend a film, play or concert? SHOW CARD M

Daily 1
Weekly 2
Monthly 3
Several times a year 4
Never 5

145. Sport

Attend or take part in a sporting event? SHOW CARD M

Daily 1
Weekly 2
Monthly 3
Several times a year 4
Never 5

146. Letter

Write letters or anything else that is more than one page in length? SHOW CARD M

Daily 1
Weekly 2
Monthly 3
Several times a year 4
Never 5

147. VTeer

Participate in voluntary or community organisations? SHOW CARD M

Daily 1
Weekly 2
Monthly 3
Several times a year 4
Never 5

148. ReadMag

Reading newspapers or magazines? SHOW CARD M

Daily 1
Weekly 2
Monthly 3
Several times a year 4
Never 5

149. ReadBook

Reading books? SHOW CARD M

Daily 1
Weekly 2
Monthly 3
Several times a year 4
Never 5

150. Listen

Listening to radio, records, tapes, cassettes or compact discs? SHOW CARD M

Daily 1
Weekly 2
Monthly 3
Several times a year 4
Never 5

151. GComput

Use a personal computer? SHOW CARD M

Daily 1

Weekly 2
Monthly 3
Several times a year 4
Never 5

152. TeleVis

How much time do your usually spend each day watching television or videos? SHOW CARD N.

Not on a daily basis 1
1 hour or less per day 2
1 to 2 hours per day 3
More than 2 hours but less than 5 4
5 or more hours per day 5
Do not have a television or video 6

153. DailyNew

Which of the following materials do you currently have in your home?

Daily newspaper?

Yes 1
No 2

154. WeekNew

Weekly newspaper/magazines?

Yes 1
No 2

155. Books25

More than 25 books?

Yes 1
No 2

156. Encyclop

A (multi-volume) encyclopaedia?

Yes 1
No 2

157. Diction

A dictionary?

Yes 1
No 2

158. NRead

I am now going to show you a list of different parts of a newspaper. Please tell me which parts you generally read when looking at a newspaper. SHOW CARD O.
DO NOT READ A NEWSPAPER = 0.
OTHER = 17

Do not read a newspaper 0
Classified ads 1
Other advertisements 2
National/international news 3
Regional or local news 4
Sports 5
Home, fashion or health 6
Editorial page 7
Financial news or share listings 8
Comics 9
TV listings 10
Film or concert listings 11
Book, film or art reviews 12
Horoscope/stars 13
Advice column 14

Special interest sections, e.g. education, motor
   or computer section 15
Personal finance, money advice 16
Other 17

159. CurrEvnt

Would you say you follow what's going on in current events, government and public affairs...

most of the time 1
some of the time 2
only now and then 3
or hardly at all? 4

160. RNewP

Sometimes people need help from family members or friends to read and write (in English). How often do you need help from others with...

Reading newspaper articles?

Often 1
Sometimes 2
Never 3

161. RInfo

Reading information from government departments, businesses or other institutions?

Often 1
Sometimes 2
Never 3

162 RForm

Filling out forms such as applications or bank deposit slips?

Often 1
Sometimes 2
Never 3

163. RInMed

Reading instructions such as on a medicine bottle?

Often 1
Sometimes 2
Never 3

164. RInGood

Reading instructions on 'packaged' goods in shops or supermarkets?

Often 1
Sometimes 2
Never 3

165. BasicM

Doing basic arithmetic, that is, adding, subtracting, multiplying and dividing?

Often 1
Sometimes 2
Never 3

166. WNote

Writing notes and letters?

Often 1
Sometimes 2
Never 3

167. RSkill

How would you rate your reading skills in English needed in daily life? SHOW CARD P.

Excellent 1
Good 2
Moderate 3
Poor 4
No opinion/not applicable DO NOT PROMPT 5

168. WSkill

How would you rate your writing skills in English needed in daily life? SHOW CARD P.

Excellent 1
Good 2
Moderate 3
Poor 4
No opinion/not applicable DO NOT PROMPT 5

169. MSkill

How would you rate your mathematical skills needed in daily life? SHOW CARD P.

Excellent 1
Good 2
Moderate 3
Poor 4
No opinion/not applicable DO NOT PROMPT 5

170. RWSatis

All things considered, how satisfied are you with your reading and writing skills in English? Are you...

excellent 1
good 2
moderate 3
poor? 4
No opinion/not applicable DO NOT PROMPT 5

171. Eye

Have you ever had eye/visual trouble of the kind that is not corrected by glasses?

Yes 1
No 2

172. PrimSec1   APPLIES IF respondent ever had eye trouble (Eye=1)

Did you have this problem while you were in primary or secondary school?

Yes 1
No 2

173. ProbNow1   APPLIES IF respondent ever had eye trouble (Eye=1)

Do you have this problem now?

Yes 1
No 2

174. Hear

Have you ever had hearing problems?

Yes 1
No 2

175. PrimSec2 APPLIES IF respondent ever had hearing problems (Hear=1)

Did you have this problem while you were in primary or secondary school?

Yes 1
No 2

176. ProbNow2   APPLIES IF respondent ever had hearing problems (Hear=1)

Do you have this problem now?

Yes 1
No 2

177. Speech

Have you ever had a speech disability?

Yes 1
No 2

178. PrimSec3   APPLIES IF respondent ever had a speech disability (Speech=1)

Did you have this problem while you were in primary or secondary school?

Yes 1
No 2

179. ProbNow3   APPLIES IF respondent ever had a speech disability (Speech=1)

Do you have this problem now?

Yes 1
No 2

180. Learn

Have you ever had a learning disability?

Yes 1
No 2

181. PrimSec4   APPLIES IF respondent ever had a learning disability (Learn=1)

Did you have this problem while you were in primary or secondary school?

Yes 1
No 2

182. ProbNow4   APPLIES IF respondent ever had a learning disability (Learn=1)

Do you have this problem now?

Yes 1
No 2

183. Mth6Prob

Have you ever had any other disability or health problem of six months or more?

Yes 1
No 2

184. PrimSec5 APPLIES IF respondent had any other health problem of six months or more (Mth6Prob=1)

Did you have this problem while you were in primary or secondary school?

Yes 1
No 2

185. ProbNow5   APPLIES IF respondent had any other health problem of six months or more (Mth6Prob=1)

Do you have this problem now?

Yes 1
No 2

Family Literacy

186. Parent

ASK OR RECORD
Are you the parent or guardian of any children aged 6 to 15 that are presently living with you?

Yes 1
No 2

187. NChild   APPLIES IF respondent is a parent of any children aged 6 to 15 (Parent=1).

ASK OR RECORD
How many children aged 6 to 15 are presently living with you?

188. ChName   APPLIES IF respondent is a parent of any children aged 6 to 15 (Parent=1).

What is the name of your youngest child, aged 6 to 15?

189. ChSex   APPLIES IF respondent is a parent of any children aged 6 to 15 (Parent=1).

ASK OR RECORD
What sex is (name)?

Male 1
Female 2

190. CAge   APPLIES IF respondent is a parent of any children aged 6 to 15 (Parent=1).

What was (name)'s age at last birthday?

191. CSch   APPLIES IF respondent is a parent of any children aged 6 to 15 (Parent=1).

How many years of school has (name) COMPLETED (not counting repeated years at the same level)?

192. Pleas   APPLIES IF respondent is a parent of any children aged 6 to 15 (Parent=1).

How often would you say (name) reads for pleasure? Would you say...

every day 1
a few times a week 2
several times a month 3
a few times a month 4
once a month or less? 5
Never (NO PROMPT) 6
Don't know (NO PROMPT) 7
Not applicable (NO PROMPT) 8

193. ReadB   APPLIES IF respondent is a parent of any children aged 6 to 15 (Parent=1).

When (name) reads, where does he/she get books?

CODE ALL THAT APPLY

Parent buys 1
Parent borrows from a friend 2
Child buys 3
Child borrows from a friend 4

Public library 5
School library 6
Gifts 7
From brothers/sisters 8
Other 9
Don't know 10

194. SatR   APPLIES IF respondent is a parent of any children aged 6 to 15 (Parent=1).

Given (name)'s age, how satisfied are you with the way he/she reads?

Would you say that you are...

very satisfied 1
somewhat satisfied 2
somewhat dissatisfied 3
very dissatisfied? 4

195. IntStm   APPLIES IF respondent is a parent of any children aged 6 to 15 (Parent=1).

The next questions will deal with your entire household. Could you please tell me if each of the following statements are true or false of your household?

196. SRead   APPLIES IF respondent is a parent of any children aged 6 to 15 (Parent=1).

Your children often see you or your partner/spouse reading.

True 1
False 2
Don't know 3

197. ChYear1 APPLIES IF respondent is a parent of any children aged 6 to 15 (Parent=1).

Your children learned to read before starting school.

True 1
False 2
Don't know 3

198. ChTime   APPLIES IF respondent is a parent of any children aged 6 to 15 (Parent=1).

Your children have a certain amount of time set aside each day for reading at home.

True 1
False 2
Don't know 3

199. ChTimeTV APPLIES IF respondent is a parent of any children aged 6 to 15 (Parent=1).

Your children are limited in the amount of time you allow them to watch TV.

True 1
False 2
Don't know 3

Parental Information

200. MothInt

The next few questions are about your mother (female guardian). Can you answer some questions about her?

Yes 1
No 2

201. MAgeSch APPLIES IF respondent can answer questions about their mother (MothInt=1)

What age did your mother (female guardian) finish her continuous full-time education? NEVER HAD=97.

202. MSchool APPLIES IF respondent can answer questions about their mother (MothInt=1)

What was the highest level of schooling that your mother (female guardian) completed?

No Schooling 1
Below Primary School 2
Completed Primary School (current leaving age 11) 3
Secondary School (current leaving age 16, GCSE
   or equivalent) 4
Upper Secondary School (current leaving age 18,
GCE A-level or equivalent) 5
Higher/Further Education NOT leading to a degree
   (e.g. HNC/HND) 6
University/College first degree 7
Postgraduate University degree 8
Education not definable by level 9
Don't know 10

203. FathInt

The next few questions are about your father (male guardian). Can you answer some questions about him?

Yes 1
No 2

204. FAgeSch APPLIES IF respondent can answer questions about their father (FathInt=1)

What age did your father (male guardian) finish his continuous full-time education? NEVER HAD=97.

205. FSchool APPLIES IF respondent can answer questions about their father (FathInt=1)

What was the highest level of schooling that your father (male guardian) completed?

No Schooling 1
Below Primary School 2
Completed Primary School (normal age 11 3
Secondary School (current leaving age 16, GCSE
   or equivalent 4
Upper Secondary School (current leaving age 18,
GCE A-level or equivalent 5
Higher/Further Education NOT leading to a
   degree (e.g. HNC/HND 6
University/College first degree 7
Postgraduate University degree 8
Education not definable by level 9
Don't know 10

Immigration

206. ImmInt APPLIES IF respondent not born in UK (COB=5)

Earlier you said that you had been born in (country). Can I just check, are you...

A British Dependent Territories Citizen,
   or a British National Overseas 1
or a full British Citizen with right of abode in the UK 2
other/don't know 3

207. YrImm APPLIES IF respondent not born in UK (COB=5)

Which year did you first come to live in this country?
(LAST 2 DIGITS OF YEAR)

208. YrsLive APPLIES IF respondent not born in UK (COB=5)

In total how many years have you lived in Britain? (LESS THAN ONE YEAR = 1)

209. SchB4 APPLIES IF respondent not born in UK (COB=5)

Before you first came to live in Britain in (year), what was the highest level of schooling you had completed?

No Schooling 1
Below Primary School 2
Completed Primary School (current leaving age 11) 3
Secondary School (current leaving age 16,
   GCSE or equivalent) 4
Upper Secondary School (current leaving age 18,
   GCE A-level or equivalent) 5
Higher/Further Education NOT leading to a degree
   (e.g. HNC/HND) 6
University/College first degree 7
Postgraduate University degree 8
Education not definable by level 9
Don't know 10

210. Ethnic

May I just check, to which of these groups do you consider you belong?
USE SHOW CARD Q.

White 1
Black - Caribbean 2
Black - African 3
Black - neither Caribbean or African 4
Indian 5
Pakistani 6
Bangladeshi 7
Chinese 8
Other 9

211. EthnO APPLIES IF respondent considers they belong to ethnic group 'Other' (Ethnic=9)

Which group (or groups) do you belong to?

Income

212. IntInc

I would now like to ask a few questions about your household. INTRODUCE HOUSEHOLD AND INCOME SECTION. (REMINDER: YOU WILL NEED PAPER SCREENER TO HELP YOU)

213. Over16

Including yourself, how many people aged 16 or over live in this household?

214. Und16

How many children under the age of 16 live in this household?

215. No1665

ASK OR RECORD FROM PAPER DOORSTEP SCREENER. THE NUMBER OF ELIGIBLE PEOPLE IN HOUSEHOLD AGED 16 TO 65?
(COLUMN E)

216. HHComp

ASK OR RECORD - HOUSEHOLD TYPE?
(REMINDER: DEPENDENT CHILDREN ARE UNDER 16, OR AGED 16-18 AND IN FULL-TIME EDUCATION, IN THE FAMILY UNIT AND LIVING IN THE HOUSEHOLD)

1 person only 1
2 or more unrelated adults 2
Married/cohabiting couple with dependent children. 3

Married/cohabiting couple with non-dependent
  children 4
Married/cohabiting couple with no children 5
Lone parent with dependent children 6
Lone parent with non-dependent children only 7
Two or more families 8

217. SrcInc

SHOW CARD R.
This card shows various possible sources of income.
Can you please tell me which kinds of income you receive?

CODE ALL THAT APPLY

Earnings from employment or self-employment.............. 1
Pension from former employer 2
State benefits 3
Interest from savings etc. 4
Other kinds of regular allowance from outside the
  household 5
Other sources e.g. rent 6
No source of income 7

218. BenList

SHOW CARD S. Which, if any, of these benefits are you
currently receiving? MARK ALL THAT APPLY.
CODE 15 = NONE OF THESE BENEFITS.

Unemployment benefit 1
Retirement pension (national insurance) or old person's
  pension 2
Child benefit 3
One parent benefit 4
Widow's pension or allowance (national insurance) 5
Disability living allowance 6
Attendance allowance 7
War disablement pension (and any related
  allowances 8
Family credit 9
Invalid care allowance 10
Income support 11
Social Fund 12
Incapacity benefit 13
Severe disablement benefit............ 14
None of these 15

219. IGross   APPLIES IF respondent has a source of income
                     (SrcInc=1-6)

Will you please look at this card and tell me which group
represents your (weekly/monthly/annual) personal income
from all these sources before deductions for income tax,
National Insurance etc.?
If you receive housing benefit please do not include it.

SHOW CARD T AND EXPLAIN ALL SOURCES MENTIONED
PREVIOUSLY ON SHOW CARD R - ENTER BAND
NUMBER.

Band 1 (up to £2704 per annum) 1
Band 2 (£2705 - £5928 per annum) 2
Band 3 (£5929 - £10400 per annum) 3
Band 4 (£10401 - £16848 per annum 4
Band 5 (£16849 or more per annum 5

Refusal 6
Don't know 7

220. IWage   APPLIES IF respondent has income from
                    earnings (SrcInc=1)

Will you please look at this card and tell me which group
represents your (weekly/monthly/annual) personal income
from only earnings from employment or self employment
before deductions for income tax, National Insurance etc.?

SHOW CARD U AND EXPLAIN - ENTER BAND NUMBER.

Band 1 (up to £4628 per annum 1
Band 2 (£4629 - £8996 per annum) 2
Band 3 (£8997 - £13000 per annum) 3
Band 4 (£13001 - £19188 per annum) 4
Band 5 (£19189 or more per annum) 5
Refusal 6
Don't know 7

221. HGross

Will you please look at this card and tell me which group
represents the (weekly/monthly/annual) income of all household members (including yourself) from all sources   before
deductions for income tax, National Insurance etc.?
If any household member receives housing benefit please do
not include it.

SHOW CARD V AND EXPLAIN - ENTER BAND NUMBER.

Band 1 (up to £5200 per annum) 1
Band 2 (£5201 - £10036 per annum) 2
Band 3 (£10037 - £17420 per annum) 3
Band 4 (£17421 - £27352 per annum) 4
Band 5 (£27353 or more per annum) 5
No income 6
Refusal 7
Don't know 8

222. TIME2

INTERVIEWER: PLEASE RECORD THE TIME.
(IN HOURS AND MINUTES - 24 HOUR CLOCK).

223. HelpBQ

INTERVIEWER: WAS ANY ASSISTANCE PROVIDED BY A
THIRD PARTY FOR THE COMPLETION OF THE BACK-
GROUND QUESTIONNAIRE?

Yes 1
No 2

Appendix D
Sample assessment items

This appendix shows a sample of the questions from the literacy assessment to illustrate some of the different tasks respondents were asked to perform. As explained in Chapter 1 section 1.5 the total number of tasks in the assessment was larger than any one individual could reasonably be asked to complete. Each respondent therefore was only asked to complete a subset of the total assessment. The assessment items were grouped into seven blocks and a Balanced Incomplete Block (BIB) design was used to arrange the blocks in different combinations into seven booklets. Each booklet contained 3 blocks of items and each block appeared at each possible location, the beginning, middle or end of a booklet in a spiral effect.

The materials used in the assessment were real items drawn from the countries that took part in the first round of the survey. They were chosen to reflect the range of different materials encountered in everyday life and each text or stimulus formed the basis for several questions which test different dimensions. For example, a newspaper article with an accompanying chart may provide material for a prose question which requires the location of information in the text, a document question which requires the location of information in the chart and a quantitative question that asks the respondent to perform a calculation. A full description of the literacy assessment can be found in Kirsch I.S. 'Literacy performance on three scales: definitions and results' in Literacy, Economy and Society: Results of the first International Adult Literacy Survey OECD and Statistics Canada 1995.

Scrambled Eggs with Tomatoes

Ingredients for 4 people:

1 garlic clove
1 onion
3 tablespoons oil
500 grams of fresh red
*   tomatoes or 500 gram*
*   can of tomatoes*
salt
1 teaspoon sugar
6 eggs

Fry chopped garlic and onion in frying pan with oil until transparent. Add tomatoes that have been peeled and chopped (if they are fresh) or mashed with a fork (if they are canned). Add salt and sugar to lessen the acidity. When the mixture begins to thicken, add the eggs, already beaten, and stir well with a wooden spoon. Cook until eggs are set.

Questions 1-3. Use the recipe for scrambled eggs on the opposite page to answer questions 1 to 3.

1. Why does the recipe call for sugar?

   _____

2. If you want to make enough scrambled eggs for six people, how many eggs should you use?

   _____

3. If you decide to make just enough scrambled eggs for two people, how many tablespoons of oil would you need?

   _____

Description of the tasks

The respondents are asked to use the recipe for scrambled eggs with tomatoes, which gives the ingredients required for four people. The first question is an example of a prose item at Level 1 where the reader is required to locate the sentence that explains 'why the recipe calls for sugar'. The second question is more complex and requires them to work out how many eggs they will need if they are making enough for six people instead of four. Here they must know how to calculate or determine the ratio needed and this is a quantitative task at Level 3. The third question asks about the amount of oil needed if the recipe is used for two people rather than four. This task was at Level 2 on the quantitative scale because a larger proportion of respondents found it easier to halve an ingredient than to increase it by 50% as in the previous question.

IMPATIENS

Like many other cultured plants, impatiens plants have a long history behind them. One of the older varieties was sure to be found on grandmother's windowsill. Nowadays, the hybrids are used in many ways in the house and garden.

Origin: The ancestors of the impatiens, *Impatiens sultani* and *Impatiens holstii*, are probably still to be found in the mountain forests of tropical East Africa and on the islands off the coast, mainly Zanzibar. The cultivated European plant received the name *Impatiens walleriana.*

Appearance: It is a herbaceous bushy plant with a height of 30 to 40 cm. The thick, fleshy stems are branched and very juicy, which means, because of the tropical origin, that the plant is sensitive to cold. The light green or white speckled leaves are pointed, elliptical, and slightly indented on the edges. The smooth leaf surfaces and the stems indicate a great need of water.

Bloom: The flowers, which come in all shades of red, appear plentifully all year long, except for the darkest months. They grow from "suckers" (in the stem's "armpit").

Assortment: Some are compact and low-growing types, about 20 to 25 cm. high, suitable for growing in pots. A variety of hybrids can be grown in pots, window boxes, or flower beds. Older varieties with taller stems add dramatic colour to flower beds.

General care: In summer, a place in the shade without direct sunlight is best; in fall and spring, half-shade is best. When placed in a bright spot during winter, the plant requires temperatures of at least 20°C; in a darker spot, a temperature of 15°C will do. When the plant is exposed to temperatures of 12-14°C, it loses its leaves and won't bloom anymore. In wet ground, the stems will rot.

Watering: The warmer and lighter the plant's location, the more water it needs. Always use water without a lot of minerals. It is not known for sure whether or not the plant needs humid air. In any case, do not spray water directly onto the leaves, which causes stains.

Feeding: Feed weekly during the growing period from March to September.

Repotting: If necessary, repot in the spring or in the summer in light soil with humus (prepacked potting soil). It is better to throw the old plants away and start cultivating new ones.

Propagating: Slip or use seeds. Seeds will germinate in ten days.

Diseases: In summer, too much sun makes the plant woody. If the air is too dry, small white flies or aphids may appear.

Questions 1-3.   Use the article about the flower impatiens on the opposite page to answer questions 1 to 3.

1.   According to the article, what do the smooth leaf surface and the stems suggest about the plant?

_____

_____

2.   Using the information in the article, list two reasons why impatiens might be considered good plants to have.

_____

_____

3.   What happens when the impatiens plant is exposed to temperatures of 14°C or below?

_____

_____

Description of the tasks

This article has two questions (1 and 3) at prose Level 2, which like the tasks at Level 1 ask the respondent to locate information. In tasks at Level 2 more varied demands are placed on the respondent in terms of the number of responses required, or in terms of the distracting information that may be present. With both questions 1 and 3 there is information that could have distracted the respondent in the sentence before the one with the correct answer.

Compound Interest
Compounded Annually

Principal £100	Period	4%	5%	6%	7%	8%	9%	10%	12%	14%	16%
	1 day	0.011	0.014	0.016	0.019	0.022	0.025	0.027	0.033	0.038	0.044
	1 week	0.077	0.096	0.115	0.134	0.153	0.173	0.192	0.230	0.268	0.307
	6 mos	2.00	2.50	3.00	3.50	4.00	4.50	5.00	6.00	7.00	8.00
	1 year	4.00	5.00	6.00	7.00	8.00	9.00	10.00	12.00	14.00	16.00
	2 years	8.16	10.25	12.36	14.49	16.64	18.81	21.00	25.44	29.96	34.56
	3 years	12.49	15.76	19.10	22.50	25.97	29.50	33.10	40.49	48.15	56.09
	4 years	16.99	21.55	26.25	31.08	36.05	41.16	46.41	57.35	68.90	81.06
	5 years	21.67	27.63	33.82	40.26	46.93	53.86	61.05	76.23	92.54	110.03
	6 years	26.53	34.01	41.85	50.07	58.69	67.71	77.16	97.38	119.50	143.64
	7 years	31.59	40.71	50.36	60.58	71.38	82.80	94.87	121.07	150.23	182.62
	8 years	36.86	47.75	59.38	71.82	85.09	99.26	114.36	147.60	185.26	227.84
	9 years	42.33	55.13	68.95	83.85	99.90	117.19	135.79	177.31	225.19	280.30
	10 years	48.02	62.89	79.08	96.72	115.89	136.74	159.37	210.58	270.72	341.14
	12 years	60.10	79.59	101.22	125.22	151.82	181.27	213.84	289.60	381.79	493.60
	15 years	80.09	107.89	139.66	175.90	217.22	264.25	317.72	447.36	613.79	826.55
	20 years	119.11	165.33	220.71	286.97	366.10	460.44	572.75	864.63	1,274.35	1,846.08

Questions 9-11. Use the table giving amounts of compound interest on the opposite page to answer questions 9 to 11.

9. You wish to invest £100 for 20 years. List all the rates on the table that will yield more than £500 in interest.

_____

10. Using the information in the table, calculate the total amount of money you will have if you invest £100 at a rate of 6% for 10 years.

_____

11. If you wanted to more than double your principal within five years, what rate of interest on this table would you need?

_____

Description of the tasks

The table gives the amount of compound interest obtained from a £100 investment over different periods of time. Question 9, at document Level 2, requires the respondent to identify the interest rates that meet the criteria given in the question. Question 11, at document Level 3, requires the respondent to identify the single interest rate that meets a complicated condition - doubling of principal within five years. Question 10 at quantitative Level 4 requires the respondent to locate the single entry in the table that meets the conditions stipulated in the question and also understand that the interest needs to be added to the principal investment in order to obtain the total amount.

Questions 4-5.   Use the advertisement for women's clothes and accessories on the opposite page to answer questions 4 and 5.

4.   Calculate the total amount you would pay on sale for a gold chain originally priced at £125 plus matching earrings originally priced at £79.

     _____

5.   You buy a bodysuit that originally cost £45 and a pair of shoes that originally cost £69. Calculate the total amount you will save on the two items because of the sale.

     _____

Description of the tasks

The advertisement gives the discounts available for different items in a sale. Question 4, at quantitative Level 3, requires the reader to locate the percentage discount for the 2 items in the question, calculate the price of the items after the discount and then add them together for a total. Question 5 involves complicated calculations where the reader must first work out what calculation is required. At Level 5 it is one of the most difficult items on the quantitative dimension.

Figures and Tables

Figures

Tables

Chapter 2

Chapter 3

Chapter 4

Chapter 5

Chapter 6

Annex tables

Chapter 2

Chapter 3

Chapter 4

Chapter 5

Chapter 6

Annex Figures

Appendix A tables

Printed in the United Kingdom for The Stationery Office
J0024129, C15, 9/97, 5673.